CASE FILES OF THE ROCKY MOUNTAIN PARANORMAL RESEARCH SOCIETY

VOLUME 2

CASE FILES OF THE ROCKY MOUNTAIN PARANORMAL RESEARCH SOCIETY

VOLUME 2

Robert Lewis and Bryan Bonner

POLYMATH
—PRESS—

Aurora, CO

Case Files of the Rocky Mountain Paranormal Research Society Volume 2
by Robert Lewis and Bryan Bonner

This book contains the true case files of the Rocky Mountain Paranormal Research Society and is intended for educational purposes, to stimulate interest and conversation concerning the paranormal, history, and science. The views and opinions expressed herein are solely those of the authors and do not necessarily reflect those of any of Rocky Mountain Paranormal's clients. Every effort has been made to ensure accuracy of the information presented, but neither the Rocky Mountain Paranormal Research Society nor Polymath Press offer any guarantees and shall not be held responsible for the use or misuse of any information contained herein.

Cover art: Bryan Bonner
Interior photographs: Bryan Bonner & Robert Lewis
Interior artwork: Spraycasso
Design and layout: Robert Lewis

First edition
September, 2024

ISBN (trade paperback): 978-1-961827-06-6
ISBN (eBook): 978-1-961827-07-3
Library of Congress Control Number: 2024944935

This volume is dedicated to the memory of James "The Amazing" Randi and Art Bell.

Though they represented opposite sides of the paranormal discussion, our work has been inspired by both.

Other Polymath Press titles by Robert Lewis & Bryan Bonner

Case Files of the Rocky Mountain Paranormal Research Society Volume 1
September, 2023

Other Polymath Press titles edited by Robert Lewis

In the Woods: A Fiction Foundry Anthology
November, 2023

Arithmophobia: An Anthology of Mathematical Horror
March, 2024

Visit www.polymathpress.com or wherever books are sold to get your copies of these and other fine books.

Praise for *Case Files of the Rocky Mountain Paranormal Research Society Volume 1*

"Forget everything you've seen on T.V. or movies about paranormal investigating – this book is the real deal, and it's smarter and scarier (and sometimes funnier). The Rocky Mountain Paranormal Research Society's case files are detailed but also exciting, written not just with a scientist's quest for truth but also a novelist's attention to pacing and background. This is a book that I suspect many of us will return to over and over in the years to come." -Lisa Morton, author of *Ghosts: A Haunted History* and *Calling the Spirits: A History of Seances*

"No flashy gimmicks here! What we have are the first-hand raw and real experiences from Colorado's finest paranormal investigators. Hands down, the most comprehensive compilation of Colorado haunts!" -Jimmy Lee Combs, director of *Terror Tales*

"Mind-blowing! The authors thoroughly dive into each case with open minds to both rational explanations and the unexplained!" -Bret Smith, co-founder of the Colorado Festival of Horror

"A thoughtful, comprehensive, and well-written compilation of some of Colorado's most interesting (and terrifying) alleged cases of the supernatural…A must-own for collectors of both the natural and supernatural histories of Colorado!" -Jack Hanley, folklorist and historian

"Gives [the subject] the respect it deserves. I am looking forward to reading future volumes as they come out!" -Jennifer Griffin, *Horror Tree*

Table of Contents

Introduction   1
   ESSAY   Ethical Considerations in Paranormal Investigation   13
PART ONE: Public Venues and Cemeteries   23
   ESSAY   The Development of Urban Legends   25
   Chapter 1   Even Scared the Ghosts Away:
               Denver Firefighters Museum   33
   Chapter 2   Ghost Lights: Silver Cliff Cemetery   53
   Chapter 3   Ghosts of Old Miners: The Phoenix Mine   63
   Chapter 4   A Trip to Deadwood: The Bullock Hotel   73
   Chapter 5   May This Room Remain Peaceful:
               Macky Auditorium   89
   Chapter 6   Smoke and Fire: Onaledge Historic Lodge   97
   Chapter 7   A Haunted Library:
               The Sarah Platt Decker Branch Library   103
   Chapter 8   A Haunted Train Station: Denver Union Station   111
   Chapter 9   The Ghost and the Teapot Dome Scandal:
               The Grant-Humphreys Mansion   119
   Chapter 10   The Other Mommy: The Spruce Lodge   131
   Chapter 11   An Investigation for the Radio:
               Heritage Square Music Hall   137
   Chapter 12   They Are All Around Us: Dunafon Castle   145
   Chapter 13   Colorado's Most Haunted Mansion:
               The Croke-Patterson Mansion   155
   Chapter 14   I'm Not Here: The Brook Forest Inn   177
PART TWO: Private Residences   191
   ESSAY   The Importance of Choosing the Right
            Paranormal Investigator   193
   Chapter 15   The Hall Ghost   199
   Chapter 16   Enough is Enough   205
   Chapter 17   The Home of Death   211
   Chapter 18   The Invisible Jellyfish   215
   Chapter 19   Had to Leave the Ghosts Behind   219
   Chapter 20   The Ghost in the Baby's Room   225
   Chapter 21   The Light Ghost   231
PART THREE: Media Analyses and Other Activities   241
   ESSAY   The Paranormal and the Press   243
   Chapter 22   The Photogenic Moth   249
   Chapter 23   The Tea Ghost   255
   Chapter 24   The Alien in the Window   261
   Chapter 25   The Denver Extraterrestrial Affairs Commission   271
   Chapter 26   The Relic of Padre Pio   289
Afterword   303
Acknowledgements   305
About the Authors   307

List of Photos and Illustrations

Figure 1.1 The Denver Firefighters Museum (Bryan Bonner)

Figure 1.2 Old Pumper 4 (Robert Lewis)

Figure 1.3 Memorial to 9/11 Firefighters (Robert Lewis)

Figure 1.4 Smoke-stained blanket (Robert Lewis)

Figure 1.5 Caleb's closet has an uneven floor (Robert Lewis)

Figure 1.6 Second floor map (RMP archives)

Figure 1.7 First floor map (RMP archives)

Figure 1.8 Basement map (RMP archives)

Figure 1.9 Could sound carry over from this attached building? (Robert Lewis)

Figure 1.10 Control objects before (left) and after (Robert Lewis)

Figure 1.11 SLS image showing a false positive for ghosts (Robert Lewis)

Figure 2.1 Silver Cliff Cemetery (Bryan Bonner)

Figure 2.2 Monument to victims of the Bull-Domingo Mine disaster (Bryan Bonner)

Figure 3.1 A Phoenix Gold Mine shaft (Bryan Bonner)

Figure 3.2 Evidence of the collapse can still be seen (Bryan Bonner)

Figure 3.3 The Phoenix Mine skylight (Bryan Bonner)

Figure 4.1 The Bullock Hotel (Bryan Bonner)

Figure 4.2 The haunted mirror with its arcane symbol (Bryan Bonner)

Figure 4.3 Our base camp in the casino (RMP archives)

Figure 4.4 The haunted coffee machine (Bryan Bonner)

Figure 4.5 The haunted balls on the bar (RMP archives)

Figure 4.6 The ball that rolled off the chair (RMP archives)

Figure 5.1 The Macky Auditorium (Bryan Bonner)

Figure 5.2 The organ room as seen during our investigation (Bryan Bonner)

Figure 5.3 The Elaura Jaquette memorial graffito (Bryan Bonner)

Figure 6.1 Onaledge Historic Lodge (Bryan Bonner)

Figure 6.2. Recreation of flash paper image (Robert Lewis)

Figure 7.1 The Sarah Platt Decker Branch Library (Bryan Bonner)

Figure 7.2 The reading room with the fireplace and painting (Bryan Bonner)

Figure 7.3 The basement with the piano (Bryan Bonner)

Figure 8.1 Denver Union Station (Bryan Bonner)

Figure 9.1 The Grant-Humphreys Mansion (Bryan Bonner)

Figure 9.2 The Grant-Humphreys theatre (Bryan Bonner)

Figure 10.1 The Spruce Lodge (Bryan Bonner)

Figure 10.2 The haunted ball in the kitchen (RMP archives)

Figure 11.1 Heritage Square Music Hall (Bryan Bonner)

Figure 11.2 The stage, ready for our radio interview (Bryan Bonner)

Figure 12.1 Dunafon Castle (Bryan Bonner)

Figure 12.2 A Dunafon Castle stairway (Bryan Bonner)

Figure 12.3 The wine cellar (Bryan Bonner)

Figure 13.1 The Croke-Patterson Mansion (Bryan Bonner)

Figure 13.2 A crumbling (now repaired) façade (Bryan Bonner)

Figure 13.3 No infant remains were found within these walls (Bryan Bonner)

Figure 13.4 The turret window from which a guard dog was allegedly defenestrated (Bryan Bonner)

Figure 13.5 The "hangman's beam" was part of a winch system (Bryan Bonner)

Figure 14.1 The Brook Forest Inn (Bryan Bonner)

Figure 14.2 The kitchen where the ghost was seen (Bryan Bonner)

Figure 14.3 The Brook Forest Inn was once a computer school (Bryan Bonner)

Figure 14.4 Did a ghostly party occupy this bar? (Bryan Bonner)

Figure 15.1 Artist's rendition of the child playing with the ghost (Spraycasso)

Figure 16.1 Artist's rendition of the ghost in the attic (Spraycasso)

Figure 17.1 Artist's rendition of the Home of Death (Spraycasso)

Figure 18.1 Artist's rendition of the Invisible Jellyfish (Spraycasso)

Figure 19.1 Artist's rendition of the ghost in the kitchen window (Spraycasso)

Figure 20.1 Artist's rendition of the ghost in the baby's room (Spraycasso)

Figure 20.2 Screen captures from the crawlspace video before (left) and after the camera tipped (RMP archives)

Figure 21.1 Artist's rendition of the Light Ghost (Spraycasso)

Figure 21.2 The light that turned itself off (left) and on (RMP archives)

Figure 22.1 The Photogenic Moth (from video provided by the client)

Figure 23.1. Our re-creation of the Tea Ghost video (RMP archives)

Figure 24.1 Rocky Mountain Paranormal's own alien in the window (RMP archives)

Figure 24.2 The alien reaches the second-floor window (RMP archives)

Figure 25.1 Mission for Inhibiting Bureaucracy (MIB) logo (RMP archives)

Figure 26.1 The relic of Padre Pio, with third class relics (Bryan Bonner)

Introduction

This second volume in the *Case Files of the Rocky Mountain Paranormal Research Society* series represents the continuation of work begun most directly about one year before we set out to write these words but indirectly a quarter of a century ago. The work continues to this day, involving not only the efforts it takes to assemble these case files into a complete and coherent whole in these books (representing our final (?) word on each case) but the Sisyphean effort involved in conducting the investigations in the first place. Both enterprises require more than full-time attention from all of our members (itself a remarkable thing as we all maintain external employment as well), but we wouldn't have it any other way. Nothing worth doing is ever easy.

In the year since we wrote our first volume, our work has continued with a specific focus on getting these books finished and into the hands of readers as quickly as possible. Our ongoing investigative activities—halted almost entirely during the Covid-19 pandemic of 2020 through...whatever year one chooses to mark as the end of that particular episode—have been gradually building back up to pre-pandemic levels. As we write these words, we're currently engaged in a new initiative to get those investigations back up to speed once again. Likely by the time you read this, we'll be back on track and investigating paranormal claims, hopefully as fast as we can write these books.

That's not to say we're in any danger of running out of material for this series any time soon. Our inaugural volume contained twenty-seven chapters representing thirteen (or perhaps thirteen and a half, depending on how you choose to count them) investigations of public venues, nine private residences, and four "other activities." The volume in your hands now contains twenty-six chapters representing fourteen public venues, seven private residences, and five "other activities." As was the case in the previous volume (and as we plan to continue in the future), the chapters will be divided into those three major sections and interspersed with several essays that we hope will help the reader develop a more complete picture of our philosophy and investigative methodology. A third volume has already been completely outlined with a similar number of entries, and we're well on our way to finalizing a table of contents for a fourth. All this even as we continue adding to our case files.

While planning those future activities, we began compiling a list of paranormal claims we'd like to look into. These include allegations of hauntings (specifically of public venues because while we are always eager to look into private cases as well, we only learn of them when people reach out to us), claims of cryptid creatures and aliens, reports of miraculous (or, conversely, demonic) activity, experiments we'd like to perform, and more. As of this writing, that to-do list has almost reached book length itself. For all the time we've been at this, we haven't even begun to scratch the surface of all the paranormal claims out there. Even in our home state of Colorado (where most of our work takes place, though we're certainly willing to travel as far as our pocketbooks will take us), there are more paranormal claims than we (or anyone) could ever hope to examine in multiple lifetimes. We're doing our best to get to as many as we can. Unfortunately, not many others are engaged in our same brand of skeptical but open-minded investigation, so we want to shed as much light as we can on as many cases as possible in the hopes both of educating the public and of discovering whatever truth is out there to be discovered.

We assume most readers of this volume have likely already read *Case Files of the Rocky Mountain Paranormal Research Society Volume 1*. However, this is not required. These books are intended to stand alone and can be read in any order. To those of you who have read the first volume, we beg your indulgence for a moment as we explain a bit of our philosophy, methodology, and the organization of these books.

With regard to organization, the first thing you need to know is that these case files are not presented in chronological order, either within this volume or within the series as a whole. This volume contains both some of our oldest and most recent investigations. The reason we don't present them chronologically is two-fold. First, some of these cases span multiple years of work and significantly overlap, so it's difficult to assign a specific date to each. In fact, the case you'll read about in Chapter 14 involved not merely years of multiple visits but more than a *decade* of research (and we're hoping to get back in to continue the investigation). Second, there are ebbs and flows in our activity, and we wanted to arrange these cases into an order that makes for an interesting reading experience. Had we done things chronologically, you'd be alternately overwhelmed with some of our most exciting cases all at once and forced to endure ones in which nothing seemed to go right. We think they're all interesting in their own way, but it's more palatable to be able to read them in a more artful order.

However, while we mixed and matched the order in which we presented our case files, we have maintained *some* order in how they are presented. Specifically, we've divided each volume in this series into three sections. The first contains cases related to public venues and cemeteries. That doesn't mean they're technically "public" in the legal sense. Many of them are privately owned. But they are "public" in the sense that they're open to the public (at least in theory; some may require an appointment or special permission to visit), they've achieved some public notoriety, and there is a much higher probability of finding publicly available historic records. Our second section contains our investigations of private residences. Though these cases seldom make the newspapers, they represent nearly half of our investigative activity, conducted as often-terrified individuals reach out to us for help in understanding some unusual thing that's happening at their homes or to their families. We think these cases are often at least as interesting as the public ones, but we're usually less able to write a full history of the location. To preserve our clients' anonymity in these cases, names, dates, and precise locations are altered or omitted, but the remainder of the stories are presented exactly as they occurred and in as much detail as we were able to discover. Finally, our third section is a sort of grab bag of "Media Analyses and Other Activities." These are the cases that don't fit anywhere else. Often, they involve a certain degree of armchair sleuthing as we've applied forensic techniques to cases reported in the media. Others have involved experiments we've performed outside the context of "normal" investigations. One case in this volume even involved the formation of a political committee.[1] Really.

Though this book (and all of our case files, if we're honest) are a bit on the ghost-heavy side, we're committed to investigating *all* sorts of paranormal claims. What's included in that category? Honestly, a lot. An essay, "Defining the Paranormal," (contained in Volume 1 of this series) sets out our answer in much more detail, but the short answer is: if it's weird, we want to know about it. Common claims include ghosts, cryptids (like Bigfoot, the Loch Ness Monster, or Mothman), claims of miracles, demonic activity or possession, aliens or UFOs, psychic abilities, and so much more. We choose not to limit our definition of the paranormal too strictly so we're able to involve ourselves in as many cases as possible. Indeed, that's one of the things that we think sets us apart from many others in this field (in addition to our philosophy, decades of experience, professional backgrounds, and investigative methodologies): while most people are *only* interested in some specific subset of the paranormal (ghosts, aliens, cryptids, etc.), we're equally intrigued by and willing to investigate *all* of them.

Because we think our philosophy and methods are what set us apart from the vast majority of other paranormal investigation teams (believers and skeptics alike), we'd like to conclude this introduction with a couple brief notes on each of those. Both of the following sections are adapted from Volume 1 of this series, with only minor updates or modifications.

The Rocky Mountain Paranormal Philosophy

There seem to be two major factions when it comes to paranormal claims. On the one hand, you have the believers who tend to seek confirmation of the existence of the ghosts/aliens/entities in which they believe. Counted on the

---

1   Though we'll explain ourselves fully in the chapter in question, it's worthy of note here that we're dedicated to political neutrality. Our members have included members of all manner of political and religious orientations, and we're committed to respect of our clients' own opinions and beliefs. However, in the case in question, a political action specifically involved a paranormal claim and we formed a committee to respond accordingly, without regard to any of the players' other political beliefs or affiliations.

other hand are the self-styled skeptics comprised primarily of non-believers who seek to debunk the paranormal claims. Rocky Mountain Paranormal proudly sits right between these two groups. When we happen to be in a contrarian sort of mood, we like to joke that we anger the believers and non-believers equally. In truth, though, we'd like to think we actually fit in rather well with both of these groups because our philosophy is compatible with what both sides at least *claim* to be doing.

One way to think about our overall philosophy is to distinguish between skepticism and what we can call cynicism. In this context, cynicism refers to rigid adherence to a foregone conclusion and not to Cynicism as in the ancient Greek philosophy pioneered by such thinkers as Antisthenes, Diogenes, and Crates of Thebes. In our sense of the word, "cynic" is often used as a derogatory term for those who disbelieve in whatever paranormal claim is being discussed. And sometimes the label applies. However, we maintain that there are cynics on both sides of the paranormal belief spectrum.

The cynical believer is one who absolutely believes in the paranormal claim in question and no amount of disconfirming evidence can change this individual's mind. Similarly, the cynical non-believer is one who absolutely does *not* believe in the paranormal claim and no amount of confirming evidence will ever be satisfactory. In the extremes, these individuals are rare. Most people, upon considering a certain weight of evidence, are capable of changing their minds. However, slightly less to the extremes of the spectrum exist a lot of people who may be in theory *capable* of changing their minds but certainly don't seem very interested in giving alternative explanations a fair and open hearing.

True skepticism requires us to treat each case from a completely neutral point of view. That's not to say we need to disregard everything we know from prior experience, but we do need to consider the possibility that any given claim may or may not turn out to be an example of a genuine paranormal phenomenon (however we choose to define that word).

An example seems to be in order, so let's consider a simple ghost story. Imagine Building X is supposed to be haunted by the spirit of someone who was murdered in the building some fifty years ago. The believer tends to want the story to be true, so even the slightest anomalous experience will be taken as proof positive that the entire story is true. The disbeliever tends to reject even the possibility of a genuine ghost or haunting and so considers even the slightest discontinuity in the story as proof positive that the whole thing is false. The true skeptic, on the other hand, takes a decidedly different approach and says: maybe the story is true, maybe the story is false, and we need to investigate (as impartially as humanly possible) to find out where the truth lies.

Rocky Mountain Paranormal insists upon the skeptical approach. We're open to the possibility of a ghost. We even hope that it's true. But we're going to look for any potential naturalistic explanations first. In our hypothetical example, one thing we'd certainly want to do is verify whether the alleged murder even took place. If we confirm it, that doesn't prove the ghost story; if we debunk it, it certainly means at the very least *that part* of the story was wrong. We'll also examine Building X and try to account for every phenomenon people have witnessed therein.

And it goes even further than that. Imagine that, on one of our investigations, we're able to explain every portion of a claim through purely natural and mundane mechanisms. Does that mean ghosts don't exist? Hardly! It's always possible the next investigation could be a real ghost. Indeed, it doesn't even disprove the existence of the ghost of a murder victim at Building X. All it means is that, of the various phenomena once thought to be evidence of the ghost, they all turned out to be caused by something else.

The late great magician and skeptic James Randi (to whom, along with the decidedly less skeptical Art Bell, this volume is dedicated) once gave another great example. He imagined trying to investigate whether or not Santa Claus could use flying reindeer as a mode of transportation. As an experiment, he imagined taking a group of, say, 100 reindeer to the top of a skyscraper and, one by one, pushing them off. If one manages to fly, well, that's case closed. But what happens if, as seems so much more likely, they all instead fall to their rather messy deaths on the street below? Does that prove that reindeer cannot fly? It does not. It merely proves that our particular group of 100 reindeer couldn't fly *or chose not to*.[2]

Does that mean we're cursed to never know anything? If we're looking for absolute certainty, the answer is pretty close to "yes." Outside of pure mathematics, absolute certainty doesn't really exist. Scientists tend not to speak of "proof," but of evidence and the strength thereof. Even in courts of law, where we do use the word "proof," we only insist upon proof beyond a *reasonable* doubt, not beyond *all* doubt, because there's always room for some doubt. What, then, are we to make of something like the reindeer experiment? We can't claim absolute proof, particularly of a negative, but we can say that the outcome of the experiment seems to make it *less likely* that flying reindeer exist. We have to remain open to the possibility that they exist out there, somewhere.[3] But if we keep experimenting often enough, eventually we keep moving that dial closer and closer to the "disbelief" side. Just as long as we never claim to move it all the way, we should be in good philosophical shape.

What about evidence in favor of something like a ghost? Imagine we watch an inanimate object move, seemingly of its own volition. Occurrences of exactly this type, in fact, will be discussed in the pages to follow. Imagine further that we do our due diligence and rule out all the natural explanations we can think of (which might include hoaxes, a draft in the room, optical illusions, hallucinations, videographic anomalies, and the list goes on). Does that prove that a ghost exists and has moved the object? It does not. The more natural explanations we rule out, the more we might move our dial toward the "belief" side of the spectrum. But without the kind of absolute proof that simply doesn't seem to exist, we can never move it all the way.

An important thing to remember is that, just because we maintain that two possible interpretations are both possible, we don't necessarily have to consider

---

2   Randi, J. (1992). *Pseudoscience and the Paranormal: Inaugural Skeptics Society Lecture* [DVD]. The Skeptics Society Distinguished Lecture Series.

3   Indeed, a believer in Santa Claus might object to our experiment claiming that of course *those* reindeer didn't fly because only those in the employ of Father Christmas have that particular ability.

them equally likely. This isn't going to turn into a lecture on inferential statistics (one of your authors has written such documents before and hopefully they'll be published soon), but that's essentially the whole point of statistical testing. We may not be able to remove all doubt, but we can at least use sophisticated mathematical techniques to quantify exactly how uncertain we are.

When it comes to paranormal investigation, there aren't a whole lot of statistical tests to be done. The kinds of evidence obtained during paranormal investigations are often qualitative rather than quantitative. We strive to obtain quantitative results as often as possible—measuring things is a key part of science—but the nature of paranormal claims frequently precludes the possibility. Nevertheless, even if it must be done qualitatively or informally, that gradual adjustment of the belief/disbelief ratio is how we gradually move toward truth.

Along the same lines, it's necessary to think not in terms of certainties, but in *degrees* of certainty. For example, we can be 100% certain of a mathematical theorem that has been proved.[4] Similarly, we can each at least arguably be absolutely certain of our own existence (that's Descartes' *"cogito ergo sum,"* or "I think therefore I am"). We can be a little less certain, but still as close to certainty as any sane person should ever require, of the existence of things like objective reality and the physical objects that surround us. We can also be reasonably assured that the sun will rise and set on schedule tomorrow. And then we get to other things that are substantially less certain, but we nevertheless seem to take for granted: that we won't die in our sleep tonight, that our car will start tomorrow morning. And so on, all the way down to things we can be pretty certain aren't the case: flying reindeer seem like a good example, as would the idea that the Earth is flat (which we would consider a paranormal claim, to be sure, but which we have thus far felt is beneath our dignity to investigate as one of our case files).

Paranormal claims, too, exist on a spectrum of plausibility. Extraterrestrial aliens make a great example. Given the immensity of the universe, many people are fairly comfortable with the idea that it seems likely for there to be some extraterrestrial life out there somewhere. It's much less likely that they've visited us on Earth. And is even less likely that they're shapeshifters that have secretly infiltrated our governments. That's not just to pick on those particular claims for being less plausible than others. It's a mathematical necessity that each successive claim in this chain is less likely than its predecessor. For aliens to have infiltrated our government, they must also have visited Earth. To have visited Earth, they must exist. By mathematical necessity, the probability of two things both being true is less than the probability of either one of them being true. This is especially easy to see when one claim is a subset of the other. Mathematical thinking like this, however, does not come naturally to the human mind. One requires substantial training to think in terms of mathematical logic.

Even when we don't have mathematical reason to say that one paranormal

---

4   At least within the constraints of our axiomatic system of mathematics—though if you really want to tumble down a mathematical rabbit hole, we suggest you should read up on Gödel's incompleteness theorems, which exposed the limitations of that axiomatic system. Be forewarned: such reading will, as the Oracle told Neo in *The Matrix*, really bake your noodle.

claim is more or less likely than another, there still seems to be that spectrum of plausibility. In our estimation, a cryptid like Bigfoot seems more likely to exist than a ghost (that's just our estimation—you're free to estimate otherwise). It's entirely possible that ghosts could exist and Bigfoot not, but given the spotty data, we estimate the best we can.[5]

Regardless of those estimates, though, you will never find any of us claiming knowledge we do not possess. Even if we might think a particular claim seems unlikely to be true, we will do our best to approach our investigation from a position of neutrality. And to the extent that true neutrality is probably impossible for the human mind to achieve, we're very careful to build safeguards against bias directly into our methods, so we can at least maintain methodological neutrality even if our attempts at philosophical neutrality fail.

At some point, though, we need to make decisions about what to believe or what not to believe. Though we will always keep investigating as much as possible, when the time comes to reach a tentative conclusion, the best approach is to apply Ockham's Razor (also sometimes spelled Occam's Razor and also known as the principle of parsimony). This is a philosophical precept which tells us that, all else being equal, the simplest explanation is usually the correct one. This is often misunderstood. "Simplest," in this context, doesn't mean "easiest to understand." It means, instead, that the explanation we should consider more likely is the one that relies on the fewest unproven assumptions. Another way to think about it is that we should always try to choose the path from Idea A to Idea B with the fewest possible intermediate steps.

Because paranormal claims, by most definitions anyway, include the existence of unproved entities, Ockham's Razor tells us we should be reluctant to accept them. When mundane or naturalistic explanations exist, they are *more likely* to be the correct ones.

All of that having been said, and while keeping all of that skeptical philosophy firmly in place, there's another piece to the Rocky Mountain Paranormal philosophy that's a bit friendlier to the believers' side of things: we really *want* at least some of these claims to be true. A big part of the reason we look into paranormal claims is in the hope that someday we might be able to confirm one of them. We may not think it's particularly likely (again, depending on how "out there" the particular claim might be), but that doesn't mean we aren't hoping.

We're horror fans through and through. There are few things in the world that bring us more joy than a good ghost story or tale of alien abduction. Even if we're somehow able to conclusively disprove one of the claims (as we have done in some but not all of the case files you're about to read), we still love the stories themselves. They're a part of our collective mythology—perhaps even part of what Jung referred to as the collective unconscious. And they're just plain fun.

---

5   We base this particular estimate on the idea that Bigfoot, while certainly paranormal (as in, not of the normal), is not necessarily thought to be supernatural. The idea of an undiscovered animal—even one as remarkable as a Sasquatch—seems to us more facially plausible than the idea of the spirit of a dead person returning to Earth. On the other hand, belief in ghosts seems to be more popular in the United States than belief in Bigfoot, so we remain open to either possibility.

The world is a big place, and undoubtedly there are still phenomena our science has yet to discover. Combine that with the number of people reporting paranormal experiences, and there's a strong temptation to think there must be something to all of this. When you add that to our predisposition to want these claims to be true, it would be very easy for us to abandon our skepticism and take even a whisper of evidence as proof of the claim. So, we're constantly walking that tightrope between open-mindedness and skepticism.

It's probably true, as Shakespeare reminded us, that there are more things in Heaven and Earth than are dreamt of in your, my, or anyone's philosophy. And with as many people as there are in the world, even if some paranormal event is exceedingly rare, there's a possibility that someone (perhaps a lot of different someones) may have experienced it. On the other hand, widespread belief in or experience of a claim does not necessarily mean it's true. Human senses are flawed and easy to deceive, and the human brain is remarkably bad at evaluating claims objectively. Paranormal investigation, properly done, is a balancing act between these competing ideas.

Beyond the question of the legitimacy of paranormal claims, though, is another piece of our philosophy: the storytelling piece. We take the position that, even if all the paranormal claims turn out to be false, they're still stories worth telling, because our stories tell us a lot about who we are. Not only are they fun, but they're often psychologically deep at a metaphorical or symbolic level.

And then there's the history. As the saying goes, those who don't learn from history are doomed to repeat it (unfortunately, someone once pointed out to us that those who do learn from history are often doomed to watch while everyone else repeats it, as history itself will often attest). Alas, a lot of our history is being lost. Wonderful historic buildings are constantly being torn down, either due to disrepair or to make room for some new development. Some of the buildings described in this series are no longer standing, making these case files a kind of ghost story in and of themselves.

Perhaps even worse, people don't take the time to read the stories of the past.[6] To many people, the study of history seems quaint or irrelevant. We maintain—no, we *insist*—this is not the case. Not only does our history provide the context for why things are the way they are today, not only does our history provide lessons for how we should proceed into the future, not only does history tell us about our own heritage, both individually and collectively, but the stories from history are the very voices of our ancestors. If you want to talk about ghosts or speaking with the dead, you really need to talk about history. Ghosts as literal entities may or may not exist—we continue to take our neutral stance—but the ghosts of the past can and do speak directly to us through our historic stories, documents, and buildings, if only we would take the time to listen to them.

Particularly when it comes to ghost stories, paranormal investigation offers a window not just into the fun and spooky tales of the ghosts themselves, but into

---

6   If you're looking for a depressing exercise, try watching some of those videos circulating around the Internet in which reporters ask random people on the street simple questions about basic history. Nothing could make a stronger case in favor of the need for improved history education.

our rich historical heritage. Part of our mission at Rocky Mountain Paranormal is to document and preserve that history to the greatest extent we can. As can read in the previous volume in this series, when a piece of history was almost lost, it was Rocky Mountain Paranormal, working in conjunction with a team of scientific experts, who managed to give a voice back to the dead in some small way (see Volume 1, Chapter 5). That's something of which we're incredibly proud, and it's something we try to do whenever we can. Both in the context of historical research and in the context of reminding people of history through our lectures (and now these books), we're using the paranormal stories as a framework to help preserve our history.

Given that paranormal stories are fun on their own and that they have such value in providing an excuse to talk about history and science, it's little wonder that some of us have found such passion in this field. And that passion is directly related to our guiding philosophy. To sum it up as succinctly as possible, our philosophy is that there's great value in paranormal stories, but they must be evaluated skeptically and scientifically to determine what is (or what is likely to be) true.

Or to put it another way: our guiding philosophy rests on equally important pillars of history, horror, storytelling, science, and skepticism.

A Brief Note on Methodology

There's no way we can provide a complete course in paranormal investigation here in this book.[7] However, it's worth taking a moment to briefly explain some of our methods so you'll understand our approach in the case files to follow. As we discuss this method, you should be aware that the precise order of events is variable. Everything always depends on the specific paranormal claim being made. But just to provide a sketch of a "typical" investigation, we're going to walk you through the steps we would ordinarily take.

In the case of a private residence, the process usually starts with a client reaching out to us, typically via email. Hopefully we're the first team they reach out to, because sometimes other teams with insufficient experience or ethics have confused or frightened the clients even more by the time the case reaches our desk. Investigations of public venues follow similar procedures, though in some cases we might have approached them with a proposal. For this brief description, we'll describe a private case but note a few key differences along the way.

Regardless of whether we're the first contact or not, we open a case and request further information from the client. Typically the claims made in these initial letters are vague enough that we don't know whether we'll be able to help or not, so we reach out to the client and request additional details. As things progress, if the case sounds like something with which we may be able to assist, we'll schedule an interview.

---

7   We've begun work on the manuscript for just such a technical volume to be published in the near future. As of this writing, we don't have a publication date yet, but depending on when you're reading these words, there's a fair chance it will already be available. Be sure to follow our own social media and newsletter and that of Polymath Press to stay up to date.

During the initial contact and interview phases, we try to collect detailed information regarding what is happening and when it's happening. We almost always request the client should keep a diary of occurrences, so we can check for any patterns. If the phenomenon always occurs on Wednesday evenings at 10:00 p.m., for instance, it wouldn't do for us to try to investigate at noon on a Saturday.

We also try to collect as much information about the clients themselves as they're willing to share. This includes any potential substance abuse problems, psychological issues, or major life changes. Additionally, we often counsel people to speak with a licensed psychological professional. This is emphatically *not* because we're calling them crazy. Though the possibility exists that people's paranormal issues could stem from deep psychological troubles, the more common reason we need to ask those questions and provide that advice is that paranormal events (whether we assume them to be genuine or psychological in nature) tend to occur during times of profound stress. Furthermore, the experience of paranormal phenomena itself can prove stressful or strain family relations, and it's important people should seek the help they need.

Once we have a detailed account of the claimed phenomena, including a diary or timeline, and possibly even including photos or videos the clients may send us, we schedule a time for an on-site investigation if we deem it beneficial. The scheduling can be difficult because we need to coordinate times that work for our team as well as for the clients, but most importantly which correspond with the timing of the phenomenon in question. Though it can be challenging, we absolutely will rearrange our schedules for these investigations when necessary.

Before we arrive at the site (and often again after leaving the site), we go through an extensive process of background research. This includes digging through newspaper archives, public records including real estate transactions, and genealogical records if they seem relevant. We spend a lot of time on the Internet, at the state archives, or at the libraries[8] finding any information that might be relevant to the case. If possible, we try to obtain blueprints of the site. If these are unavailable, we'll make our own approximation of a site map during our investigation. Background information at public venues is almost invariably easier to find and more abundant.

When we arrive at the site, we're not doing what you often see on television. First of all, we don't bring a television crew (much to the surprise of some of our clients). We also ask them not to invite guests. Though we realize our work is interesting to people, we need the site to be as controlled as possible, and that means not having a bunch of people running around contaminating data. At public venues, we try to conduct investigations after hours when only ourselves and employees are present (at least whenever possible; there are exceptions to every rule, and some venues do stay open 24/7, so we adjust as necessary). We'll then tour the building, determine locations for cameras and other monitoring equipment, set up our base camp, and prepare for the investigation. We ask the clients

---

8   To anyone interested in any kind of research (paranormal or otherwise), take note: librarians are your best friends. They don't just spend their time rearranging the bookshelves; they're almost like wizards in their ability to find whatever information you need.

to pretend we're not there to the greatest extent possible and continue with their normal routine—it is, after all, during their normal routine that the phenomenon is claimed to occur.

Most importantly, once everything is set up, we shut up for the bulk of the investigation. We don't try to carry on conversations with the ghosts or run around talking to each other. Noise contaminates data. Furthermore, if we start talking to the alleged ghosts, we'd be psychologically priming ourselves not only that there is a ghost (which is what we're trying to determine in the first place) but also to expect specific answers to our questions. We might announce something like "if there are any spirits, feel free to make yourselves known to us" (with heavy emphasis on the word "if,") but that would be the extent of it. If the particular claim necessitates a more direct interaction with the alleged entity, we are careful to always do so hypothetically.[9]

Equipment we might use on site includes still cameras, video cameras, 360-degree or spherical cameras, electromagnetic field (EMF) detectors, seismometers, thermometers, audio recorders, and any other tools that might seem relevant to a specific case. We sometimes also include air quality or other environmental monitors to alert us if the house may have something like a gas leak (which can, under some circumstances, cause hallucinations).

Because the paranormal is an ill-defined field of study, we don't know how to measure a ghost, alien, or demon. What we can do is establish baseline readings on our various instruments and then look for anomalous readings. If an anomaly should occur, we don't immediately assume it's supernatural. Instead, we note it, and follow up to see if we can find a way to explain it.

There are plenty of devices and gizmos claimed to be part of the ghost hunter's arsenal that we tend not to use most of the time. These are the devices that were specifically designed for paranormal investigation. Such devices are either hoaxes or intended for entertainment purposes rather than forensic investigation. There's no way to calibrate a device to detect a ghost since we've never even conclusively proved that ghosts exist or figured out their properties. We do own these devices and will experiment on or with them, but they're not part the gear we'll use on site unless the claim in question is specifically related to one of them.

At the end, we'll present the client with a report of our findings. This will include all of our background research as well as all of our measurements. The most important part will be our lists of findings, and they come in three categories. First, we list explainable events, or apparently paranormal phenomena we were able to explain naturally. Next, we list any phenomena or events we witnessed but have not yet been able to explain. Finally, we list any phenomena the clients reported but we were unable to witness or recreate. Along with this final

---

9   Whether or not to engage in these kinds of interactive activities on an investigation has been a source of some controversy among other investigators and much discussion within Rocky Mountain Paranormal. We maintain that there *is* a way to do this sort of thing legitimately (though we usually tend not to for the reasons indicated) but that it's difficult and requires sophistication in the study's design. A more complete treatment of the subject will appear in our forthcoming "how to" book.

list, we'll include any possible explanations we can think of, with the caveat that until we actually witness or test it, those explanations are speculative.

After we present our report, the clients often still have difficulty dealing with their issues. We're happy to help them to the extent we're able, but we always encourage them to speak to a psychological professional or work with their own clergy. Should it be necessary, we're always happy to return to conduct follow-up investigations if the anomalous phenomena continue.

You'll note that our methods differ from what you tend to see on television in which the so-called investigators run around in the dark scaring themselves and each other. We mark a distinction between the activities of "ghost hunting" and "paranormal investigation." Ghost hunting is what you see on TV. It's a lot of fun, but it doesn't yield quality data. Paranormal investigation is methodical, scientific, and often quite boring or tedious to do.

Ghost hunting does have its place. It actually has two purposes. The first is purely for entertainment. As horror fans, we can appreciate running around in the dark and scaring ourselves silly. We just don't claim that to be part of a scientific investigation, no matter how many beeping gizmos we carry with us when we do so. The second is that ghost hunting can actually precede a proper scientific investigation and can be valuable in that context. It gives the phenomenon a chance to manifest itself under less stringent controls. The important thing to realize is that, when those controls are loosened, everything observed must be taken with a grain of salt.

In mathematics, there are two kinds of data analysis: exploratory and confirmatory. The former is essentially mining data to see what we can find. The latter is to confirm the validity or truthfulness of whatever we found in the former. Ghost hunting, in its proper context, can be thought of as a kind of exploratory analysis. Go to a haunted place and see what happens. But, as is the case in mathematics, you *must* follow it up with more rigorous confirmatory analysis.

ESSAY
Ethical Considerations in Paranormal Investigation

Political views differ on the appropriate degree of official regulatory oversight of professional activities, and Rocky Mountain Paranormal remains committed to neutrality on such political questions. Our members and colleagues may have opinions on the topic, but as an organization we don't presume all of those opinions are in agreement. However, it's worth it to take a moment to reflect upon the fact that many professions are highly regulated and have varying degrees of government or peer oversight. Lawyers must be admitted to the Bar. Physicians must be licensed by the state and are monitored by medical boards. And so forth. There is no particular body of law nor any meaningful professional oversight of paranormal investigators. What few organizations exist that do claim to offer some degree of oversight have no means to enforce their prescriptions (nor any claim to legitimacy).

We are not arguing here that legislatures should target paranormal investigators with regulatory oversight. Far from it. But we do argue that investigators must, in the absence of such regulation, develop a strong sense of professional ethics. Regulations and professional societies in other professions are not dispositive in this matter, but they can and should provide a framework of potential issues to consider. We consider the ethical guidelines of private investigators, research academics, and psychologists or therapists to be of particular interest in this field because our work in paranormal investigation often overlaps with the kinds of work in which those professionals engage. We'll not explore the ethics literature on those professions in great depth here (it's beyond the scope of this book and such documents are readily available online or at any library) but we recommend people considering a vocation in paranormal investigation should read them in more detail. Here, instead, we're going to present our own distillation of those ethical guidelines as well as some that we've developed specifically with regard to the paranormal. Hopefully this will serve as a guide for would-be investigators as well as a "peek behind the curtain" into the attention we give each of our investigations.

Clients interested in engaging the services of a paranormal investigation team or society might also consider this essay as a partial guide to what you should expect and the standards to which you should hold anyone you welcome into your home or business. An additional essay focused more specifically on that topic will follow later in this book (as the opening essay to Part Two).

Because we can't resist the temptation to add some "artistic flair" to what might otherwise be a fairly dry set of ethical precepts, we've elected to present our guidelines in the form of Biblical Commandments. No religious or anti-religious claim is intended.

1) Thou shalt report thy findings honestly.

Though we admit paranormal investigation is fringe science, it should nevertheless be treated as a legitimate science and its practitioners should behave as true scientists. The most important rule, therefore, is to respect the truth—whatever the truth might or might not turn out to be—and to report all of one's findings completely and honestly. If you're hoping to find a ghost but instead find a leaky flapper valve on a toilet (Volume 1, Chapter 23), you need to admit it and say so. On the other hand, if you're expecting to debunk everything but you hear a voice you can't explain (Volume 1, Chapter 3), you need to report that as well. The report should be detailed enough that a future investigator could attempt to replicate your findings and should account for everything you could *and* couldn't explain. Speculation as to the cause(s) of any phenomena are certainly acceptable but must be clearly labeled as speculation and presented hypothetically.

Even in legitimate science, there's a crisis in terms of reliability and replicability of many studies. A full account of that controversy is beyond the scope of this book, but part of the problem comes down to a bias in favor of positive results. Studies that *didn't* find what they were looking for tend either not to be published or relegated to lower-ranked journals. The most-read journals tend only to publish positive and noteworthy results. That becomes particularly troublesome when one realizes that, by tradition, scientific studies accept a 5% chance of a false positive result. What that means in practice, to oversimplify a serious and complicated issue, is that if twenty studies are conducted, say, to determine whether Drug X is effective or not, and only one of them finds a positive result, that outlier study is the most likely to be published, even though a look at the entire body of research suggests the possibility that the positive result may have been a false positive or a statistical anomaly.

The same thing happens in paranormal investigation, albeit less formally than in the scientific literature. We assume everyone who reads books or articles on paranormal matters is interested in the paranormal. Fair enough. But when those publications only offer the positive results, it could easily create the illusion that a particular location is haunted (or a particular UFO sighting was really an alien or whatever the claim may be) when the reality could very well be that of a hundred investigators, only one of them witnessed anything out of the ordinary. We're not saying to disbelieve the one out of a hundred, but we are saying it's necessary to listen to all hundred. In these books, you'll notice that we present *all* of our cases, not just the most exciting ones, because doing so is part of our ethical obligation to report our findings as completely and accurately as humanly possible. Sometimes that even means having the courage to speak the truth when everyone else in the room would very much prefer to hear a lie.

2) Thou shalt comport thyself as a scientist.

There's nothing ethically wrong with a non-scientific ghost hunting venture if all you mean to do is to have some spooky fun. But if you're claiming to conduct an investigation, you must behave as a proper scientist. We maintain this is not only a concern with regard to the validity of an investigator's findings but an ethical necessity. If one values truth, as one should, then only fair and open-minded scientific inquiry could hope to uncover the truth about a variety of paranormal claims. That is, in and of itself, an ethical concern for us because, like our first "commandment" it's related to our ongoing project of determining and disseminating the true facts of any given paranormal claim.

To comport oneself as a scientist means a lot of different things. What it does *not* mean is that one must obtain a doctorate from a prestigious university, obtain federal research funding, wear a lab coat (though we do occasionally like to model our stylish lab attire), or otherwise adopt any of the stereotypical trappings of the scientific enterprise. What it *does* mean is that one should state one's hypotheses as clearly as possible, that one should design tests or experiments and establish controls before collecting data, that one should remain open-minded to alternative hypotheses, that one should consult with experts when one's research takes one outside of one's own field of expertise, that one should engage in spirited debate when necessary but should always remain receptive to opposing ideas, that one should be willing to change one's mind in the face of a certain amount of evidence, that one should insist upon a certain amount of evidence rather than changing one's mind arbitrarily, that one should be cognizant of the various logical fallacies and cognitive biases to which we're all susceptible, that one should insist upon objective evidence rather than arguments from authority, and so forth. Probably the most important part of this "commandment" is that a good scientist approaches an investigation from a position of methodological (if not philosophical) neutrality and avoids taking any sides until the data are in.

To put it another way, we'd like to borrow from an article published as part of a program at the Michigan State University. It asserted that a "good scientist" should be curious, patient, courageous, detail-oriented, creative, persistent, communicative, open-minded, free of bias, a critical thinker, and a problem-solver.[1] We emphatically agree and consider it not only a matter of good practice but of professional ethics.

3) Thou shalt comport thyself as a professional.

In addition to the professional standards of scientific practice outlined

---

1   D'Augustino, T. (2017). What makes a good scientist? *Michigan State University Extension.* <https://www.canr.msu.edu/news/what_makes_a_good_scientist> (accessed July 17, 2024).

above, one should strive to behave professionally in all of one's business, especially when interacting with clients or the public. Professionalism itself can be difficult to define. On some level, we find ourselves in the same position as Justice Potter Stewart who famously said (albeit about a rather different subject), "I know it when I see it." But while we may not be able to define it perfectly, we nonetheless think it's an important ethical guideline.

A few things it assuredly does mean are pretty easy to identify, though. You shouldn't drink alcohol (at least not to excess) on an investigation. A glass of wine with dinner is probably fine, but a bottle of scotch while you're supposed to be observing a scene is probably not. You should address the clients and other individuals present with respect, *especially* when you have to ask them about personal matters. You should present yourself well. That doesn't necessarily mean suit and tie (though a bit of formality never hurts) but it does mean you shouldn't show up to someone's home or place of business looking like you just crawled out of a swamp. When you're on site, you should remember that you're there to work, and prioritize business over socialization.

None of this means you can't have fun on an investigation. We often do. But it's important to treat these investigations as professional encounters. Laugh and have a good time, but be sure to get the work done and try not to make people uncomfortable while you're doing so. People even joke and laugh momentarily during Supreme Court oral arguments of grave national significance, so we're certainly comfortable with a paranormal investigation having some levity. In fact, it's even preferable a lot of the time. But like the lawyers and Justices at the Supreme Court, you need to make sure you're still getting the work done well and efficiently.

4) Thou shalt not claim false expertise.

Admitting that the paranormal is ill-defined does not make it any less important a field of study, particularly in light of the sheer number of people not only interested in paranormal ideas but who claim personal experience with some sort of paranormal phenomenon. However, admitting that the paranormal is unproven and ill-defined *does* mean we need to be extremely cautious of any claims of expertise. Arguably, there is no such thing as an expert in the paranormal. How could one be an expert in something that has not yet been proven to exist?

But that's an oversimplification. There are domains of expertise within paranormal investigation, and we do claim a degree of expertise in certain matters when it is appropriate to do so. Our members are professionally trained experts in a number of related fields including psychology, deception detection, forensic media analysis, photography, and many more. Over the years we have also developed substantial expertise (mostly from experience rather than formal schooling in this case) in folk history and paranormal *lore*. But while it's perfectly legitimate to claim expertise (as we do) in paranormal lore, it's another thing entirely to claim expertise in *the paranormal*. To take ghosts as an example: we are not experts in ghosts. There's no such thing as an expert in ghosts, because no one has ever been able to study a ghost (if such things exist) to the degree necessary to develop what we could call an expert body of knowledge. But after more than twenty-five years of listening to and investigating claims of ghostly activity, we do claim some expertise in the *claims* surrounding ghosts.

When people reach out to us, they treat us as experts because we're the ones who spend our lives looking into these claims and have been at it for longer than some of our younger clients have been alive. Fair enough. But we simply can't answer some of their questions.

Just as a single illustrative example, a reader of our last book asked one of our members recently, "do ghosts know they're dead or not?" We don't and ethically can't claim the expertise necessary to answer such a question. No one can. We can cite the paranormal lore and explain that there are plenty of different stories and that in some of them the ghosts seem self-aware and in others they don't. That's within our expertise. But to actually answer the specific question the way it was asked? No one possesses that knowledge, and it would be unethical to pretend otherwise.

A related topic comes up when paranormal investigators are sometimes asked (and this does happen more often than you think) to engage in activities that are more properly the province of mental health professionals or clergy. On some occasions, families or individuals have asked us to settle familial disputes tangentially related to the paranormal claims. On numerous occasions, we've been asked to perform cleansings, blessings, or exorcisms of an individual or property. While we are well-read in psychology and religions, these are not activities within our professional purview and so we must recognize a duty to refer the clients to external assistance in those cases. That doesn't mean we step away. We're more than happy to help facilitate contact with the appropriate professionals. If we've built a relationship of trust with our clients (as we always aim to do), we're happy to be present when they work with those professionals (if it's appropriate to be). But we're not doctors and we're not clergy and it would be unethical to pretend otherwise.

5) Thou shalt not investigate without an invitation.

A few places out there are legitimately public. Government offices, public parks, that sort of thing. As such, members of the public do, under certain circumstances, have a genuine right to engage in investigative activities at such locations with or without explicit permission. That's a legal question and should be directed to your lawyer for further advice because there are also legitimate restrictions on one's ability to conduct potentially disruptive investigative activities even on public property.

But most of the places we investigate—and the same goes for most other investigative teams—are private. Or even if publicly owned, they're not necessarily *open* to the public. A good example of that was our investigation at the Denver County Jail (Volume 1, Chapter 10). Yeah, jails are public buildings, but they're also secure buildings. People can't just walk in and out willy nilly. That's kind of the whole point of a jail. The reason we were able to investigate there wasn't that we asserted some privilege, but that we were specifically invited to do so at a time during which the inmates had been transferred out of certain buildings. Most of the rest of our investigations are on private property of some sort or other. As such, permission is required.

Over the years, we've noticed a lot of so-called paranormal investigators have broken into locations (or at least trespassed after hours) to conduct their work. This is unethical, dangerous, and inappropriate. You *must* get the permission of the property owner before conducting an investigation. One investigator, whose identity we'll not reveal, once advised his followers not only to sneak into a closed graveyard after hours, but to carry firearms in case they encountered any guards. Now, we maintain our professional political neutrality on issues of firearms in general—some of our members like them, others don't. But this is not an example of ethical behavior. By trespassing without permission (and doing so while armed), this individual threatens the lives not only of the security guard but of his own crew. The appropriate way to gain access to a graveyard like that is to write a proposal and ask the permission of the property owner or manager. We're not going to say it's inappropriate to carry a firearm on an investigation for self-defense purposes. Some investigations take place in shady neighborhoods or deep in the woods and it makes sense to carry protection if you're comfortable with it and have the proper training. But it should be noted that carrying a firearm while engaged in otherwise criminal activity such as breaking and entering will get you in a world of legal trouble in addition to the ethical considerations.

With all that in mind, we consider ourselves akin to vampires. We never set foot on any property without permission. If a place is open to the general public, we might check it out *informally* without explicit permission (for example by simply dining at a supposedly haunted restaurant). But when it comes time to haul in all of our gear for a proper investigation, the only ethical option is to do everything above board.

6) Thou shalt not damage thy neighbor's property.

Related to the rule of getting permission before engaging in an investigation is a rule that probably should go without saying (but here we are, saying it anyway). Don't harm the persons or damage the property of others. Whether you've been explicitly invited or not (and you should be), you're a guest on the property and you should leave it as you found it. That means when you're setting up your gear, you need to be sure you don't damage anything. If you want to screw into a beam to mount a camera, make sure you get the property owner's permission to do so. If you're wandering around the woods, make sure you take your trash with you.

Unless given permission to do otherwise, always follow the cliched travel advice: leave only footprints, take only photographs. Yes, our own cabinets of curiosities are full of artifacts from our investigations, but they were taken with explicit permission from the owners or given to us as gifts. We never take anything without permission and we're always careful not to damage anything.

Accidents do happen, of course. Ethical considerations require due diligence to avoid them, but they're called accidents for a reason. If something does happen, own up to it and offer to make restitution to the client. If the costs are high, you may need to work with your insurance company or the client's, but make sure you do everything honestly and above board.

7) *Thou shalt respect the religious beliefs of thy clients.*

Paranormal claims often intersect with religious beliefs. It's just part of the game. And of course, our members all have their own religious ideas and affiliations. People we've worked with over the years have ranged from the staunchest of atheists to the most devout of believers. And that's just within our own crew. If we consider the clients with whom we've worked over the years, they've included just about every religious affiliation we can think of. Because religion and paranormal claims sometimes intersect, it is of paramount importance for the ethical investigator to remain respectful of religious beliefs.

That doesn't mean one should hide one's findings for fear it might conflict with a religious belief. If someone believes God is speaking to him but it turns out just to be a faulty radio, then demonstrable truth trumps religious deference. But to the extent that any claims remain unsolved, they should be investigated either with complete neutrality or within the context of the client's own religious beliefs. If clergy need to be called in for the emotional wellbeing of a client, said clergy should belong to a religion or denomination with which the client him/herself is comfortable, even if that denomination might be diametrically opposed to the investigator's own beliefs.

An important addendum to this commandment: respect for the client's religious beliefs often requires a familiarity with various religions of the world. No one can know everything, of course, but we think it's our duty to be well-read on a variety of religious claims and practices so we're prepared to understand, and to understand deeply, the claims each client makes. If we enter a home and see a religious altar, we try to make sure we have an understanding of its meaning for the client. We've read of unfortunate cases in which first responders saw a religious altar they didn't recognize and falsely assumed the individual was involved in some kind of Satanic practice, with sometimes disastrous results. We require our members to do better.

8) *Thou shalt not harm thy client.*

There are numerous circumstances in which a paranormal investigator could, intentionally or inadvertently, cause harm to a client. Some of those are specifically addressed in other commandments on this list (in terms of protecting their property and their privacy and so forth), but it's worth considering more subtle ways one could adversely affect a client as well.

The most obvious thing that comes to mind is that one should always tread carefully when dealing with a client who may be recently bereaved. One of the most common paranormal claims is the "death visitation" in which an individual believes he or she has been visited by the spirit of a deceased loved one. Another popular paranormal belief is that psychic mediums can communicate with the deceased. When looking into these or other related claims, investigators need to adopt a particularly strict set of ethical procedures to ensure that their activities don't alter or disrupt the grieving process or supplant genuine memories of the deceased with a fabrication. Similar considerations should be taken when working with children or the mentally impaired.

A good place to start with any investigation is to obtain informed consent before working with any client. Conduct a risk assessment based on the background information you've been provided (including both physical and emotional risks) and present this assessment along with your research plan before working with the client. Explain clearly that you can't guarantee any particular finding. Explain what methods you plan to use. Offer to refer the client to professionals if needed. Ask the client if there are any medical or psychological issues you should be aware of and develop a plan to act accordingly.

9) Thou shalt respect thy client's privacy.

Paranormal investigation, particularly when done at a private residence, can get rather personal. We have to ask clients questions they might ordinarily be reluctant to discuss even with their psychologist or clergy. If there are troubles in the family, we need to know about it because those tend to correlate with paranormal phenomena.[2] The same goes if there are problems with drugs or other forms of substance abuse in the family. Even matters like cleanliness and hygiene can play a role in an investigation as environmental factors can become relevant. Add to all of that the fact that many people feel a certain degree of stigma associated with being a paranormal claimant. Though we approach people without prejudice on that matter (as is also one of our ethical guidelines, stemming from several of the commandments already listed), not everyone is open-minded about such things. Some people could risk losing their job if their boss discovers they're claiming demonic possession or alien abduction. Paranormal claims can even affect real estate values in some cases. Whether or not such judgments are appropriate, it becomes the responsibility of the paranormal investigator to tread softly when dealing with these matters.

It might seem that this commandment is at odds with our first commandment to be as complete and detailed in our reports as possible. And at times, these can seem to come into conflict with one another. However, as a general rule, we believe that paranormal investigators should treat themselves as any other professional and should treat their case files as confidential, only publishing any findings once they have been thoroughly stripped of any potentially identifying information. It may be necessary, for example, to disclose a drug habit as part of a finding (particularly if that turns out to be the cause of the paranormal claim). But it can be done anonymously so the client is never embarrassed by the report.

Public venues are a different matter. Because they are public-facing and their paranormal claims have been reported in the press, public interest may trump concerns about privacy. But in the private cases, the identities of the client should

---

2   Skeptics would claim that the psychological disturbance can, in extreme causes, cause hallucinations or delusions. While this is true in at least some cases, the believers would counterargue that psychological disturbances can make individuals particularly receptive or vulnerable to a variety of paranormal entities. As a matter of our practiced neutrality, we acknowledge both possibilities and remain agnostic on the subject until we've had a chance to investigate more completely.

be kept confidential and separate from any published finding absent extraordinary countervailing circumstances. What are such countervailing circumstances? One would be a lawful court order to disclose such information. Unlike medical professionals, lawyers, and clergy, paranormal investigators do not benefit from an established legal doctrine of non-disclosure and are therefore likely to be subject to subpoena. To our knowledge, no one has ever tested these ideas in a court of law, and we're not offering legal advice, but that at least conceivably could be a circumstance in which privacy may be breached. Another circumstance would be when the life, health, or safety of an individual is at stake. We had a case (which will be reported in a future volume) involving such a severe case of child abuse that we felt the need to report our findings to the appropriate authorities to protect the child. But absent these kinds of extraordinary and rare circumstances, you shouldn't name your clients publicly unless they give you permission to do so.

10) Thou shalt not charge for thy services.

This is a big one. Enough people are looking for assistance with paranormal claims that one could probably carve out a rather successful business investigating those claims for a fee. We consider this *highly* unethical. In fact, we'll go as far as to say that if a paranormal investigator wants to charge you for his or her services (with the exceptions outlined below), you should immediately cease contact with that individual.

The fact of the matter is, as we keep saying, the paranormal is ill-defined and whatever expertise one might legitimately claim in the paranormal field is necessarily on the periphery of the issue. To us, it is inappropriate bordering on criminal to charge someone for ill-defined services. Paranormal investigation should be done out of curiosity, out of a desire to help people, out of passion for the field. It should not be done for a profit.

That said, we do carve out an exception for actual expenses. While we at Rocky Mountain Paranormal strive to completely self-fund our operations, we understand that some people might not have the means to do so. We certainly wouldn't object if an investigation team asked for travel expenses or for any other actual expenses incurred during the investigation. For our part, we try to not even do that if the client is local to us. The only times we've investigated on someone else's dime were occasions when we've been flown out of state to engage in our work. And in those cases, we only asked for travel, room, and board. Never a profit.

But that doesn't mean one can't ethically make a profit in this field *indirectly*. There's nothing wrong with selling your books or accepting an honorarium for speaking engagements. That's all perfectly fine and dandy (as long as you're following the rest of these commandments when you're reporting your findings). There's money to be made in this business, and we're certainly not trying to condemn ourselves or our fellow investigators to a life of poverty. We simply insist that individuals (or even businesses) should not be charged for the investigative services themselves.

As a side note, we also object to businesses (usually the famously haunted ones) that want to charge investigators exorbitant fees for the privilege of investigating on their property. While these businesses are private and such is their prerogative, it leaves a foul taste in our mouths. Just as we wouldn't charge for investigative services beyond actual expenses incurred, we also wouldn't *pay* for the privilege of doing our work beyond actual expenses, though that's less of an ethical commandment and more of a stubborn refusal on our part.

We're sure we could easily come up with at least ten more ethical commandments for paranormal research, but we think these guidelines represent a pretty good start. We also encourage members of investigation teams to self-police (because as we mentioned, there is no strong regulatory framework in this field). If someone on your team acts unethically, call them on it. If you see another team acting unethically, you may need to proceed with some delicacy, but it would benefit everyone if you took some kind of action.

PART ONE:
Public Venues and Cemeteries

While most paranormal claims are experienced and dealt with privately and seldom make the newspapers (and even if they do, tend to be quickly forgotten), there are some places that have become famous for their alleged paranormal phenomena. Many hotels and former hospitals are thought to be haunted. Same goes for plenty of cemeteries, event halls, and schools. In this section's fourteen chapters, we're going to look at some of Colorado's most famous haunts as well as some lesser-known but still fascinating settings for alleged paranormal phenomena.

ESSAY
The Development of Urban Legends

Work in paranormal investigation overlaps with a lot of different fields of study. One of the most prominent of those is folklore. After all, what are paranormal claims but a subset of our cultural heritage? That's not to say these are "mere" folklore or urban legends without any truth to them. On the contrary, even some of the most famously false urban legends have at least a kernel of truth at their heart, and paranormal claims have been neither proved nor disproved conclusively. But in order to understand how paranormal claims have developed over the years, we've needed to become somewhat expert in the development of folklore and urban legend. In this essay, we want to briefly explain how urban legends originate, evolve, and propagate themselves over time. Obviously, such a brief treatment of the subject will necessarily be somewhat superficial, as many individuals have earned doctorates in this field and make such study their life's work. No essay can compete with that body of literature. But what we can do is to at least *sketch* the process of folkloric development and explain how that process applies to the paranormal.

One of the nice things about having worked in paranormal investigation for as long as we have is that over the years, our case files have (sometimes quite accidentally) documented the entire folkloric process of certain legends, from formation through decades of continual change. This essay isn't the place to fully document those cases (they will get their own chapters in these books), but it's worth mentioning in advance that one of the cases in this volume (Chapter 5 on the Macky Auditorium) is such a case. Throughout our investigation and in the years since, we've documented not only the inciting incident that sparked the legends of a haunting but also the locals' changing views on the matter over time.

To begin with, some terminology is probably in order, especially because connecting the concepts of urban legend and paranormal lore, while we think it's quite sensible, can lead to some confusion. The phrase "urban legend" is itself something of a misnomer because not all of these legends have anything to do with urbanity. Rather, some folklorists prefer to call them "contemporary legends" because they consist of what urban legend expert and collector Jan Harold Brunvand defines as "all those bizarre, whimsical, 99 percent apocryphal, yet believable stories that are 'too good to be true'" and "the modern narratives that folklorists have collected and studied under [certain] well-known titles."[1] That definition, while accurate in the description of urban legends in general, doesn't quite fit our own needs if we want to include the paranormal. After all, there are those skeptics who will consider the "believable" part of the quotation questionable when applied to something like a ghost story; conversely, believers might object to the "99 percent apocryphal" and "too good to be true" descriptors. If we want to include the kinds of work we at Rocky Mountain Paranormal conduct, we need to expand that definition, and so we propose (and will follow for our own purposes) the following definition of an urban legend: a compelling yet unverified (sometimes unverifiable) tale of unusual happenstance, generally transmitted through social or memetic means and often evolving in the retelling. With regard to "memetic means," we'll come back to define that more precisely in a moment.

A good analogue for the development and transmission of urban legends is the "telephone game." As it's generally played among children, one individual—whom we can call the originator—comes up with some simple phrase and whispers it to the next person in line. That person listens to the message and then whispers it to the next person, and so on, until the individual at the very end of the line—whom we can call the receiver—announces the message he or she has received. In most cases—indeed, almost invariably—the message announced by the receiver bears strikingly little resemblance to the one provided by the originator. Through small evolutionary changes at each step of the transmission process, the message gets garbled and reinterpreted. Sometimes individuals introduce intentional corruptions to change the message for their own reasons, but even if everyone intends to faithfully pass on the message, slight errors in their hearing propagate down the line until the final message is irredeemably altered.

Though we can understand in principle how urban legends and other related forms of folklore propagate, it's often difficult if not impossible to trace the origins of one in particular. In his entire *Encyclopedia of Urban Legends*, Brunvand identifies only two such cases (out of hundreds included in the book).[2] However, even when definitive answers are not possible, it's often not terribly difficult to at least paint a partial picture of such an evolution.

For example, consider the stories of ghosts at the Stanley Hotel in Estes Park, CO (Volume 1, Chapter 14). Today, there are more ghost stories than even we have been able to count. The hotel offers daily ghost tours, and the cast of spirits alleged to haunt the beautiful hotel's halls must now number in the hundreds, if not thousands. We've even heard claims that misuse of a spirit board at the hotel has invited an evil (perhaps demonic) presence. Where does it all come from? As we described in the previous volume, some of the stories seem to have "naturally" occurred at the hotel. Perhaps because it's truly haunted or

1   Brunvand, J. H. (2001). *Encyclopedia of Urban Legends*. New York: W. W. Norton & Co.

2   *ibid.*

perhaps just because historical circumstances were such that the place was ripe for such stories. But it wasn't until Stephen King stayed at the hotel and reported an unusual experience (itself often magnified in the retelling) and subsequently used the hotel as the basis for his novel *The Shining* that the hotel developed a true international reputation for its hauntedness. And in the years that have followed, with ghost hunters and paranormal enthusiasts flocking to the hotel in droves, the story has morphed from the original few ghosts to the situation today. Which isn't to say none of the recent additions are legitimate. But it seems almost unthinkable that *all* of the stories are legitimate, and separating the wheat from the chaff becomes difficult as the folklore grows.

A case whose origins we were able to trace definitively was also featured in our first volume: Josephina's Restaurant (Volume 1, Chapter 2). In that investigation, we found that the ghostly lore—still repeated today as a true story—was deliberately written by a specific identifiable individual in an attempt to generate publicity (it should be noted that this individual was completely honest with us about the whole thing). Looking back further, we identified potential sources in folklore that may have inspired the story this individual wrote, but looking forward we find that people still consider the building (though no longer home to the same business) one of the haunted hotspots in the Denver area.

Though not from our own case files, the Slender Man legend is another prime example of a story whose entire history is fairly well documented. This legend tells of a supernatural entity—Slender Man (or sometimes rendered in a single word as Slenderman)—characterized by his lanky girth, unnatural height, featureless face, and black suit and tie. He is a decidedly creepy character whose very appearance telegraphs his malevolent intentions. In many versions of his stories, he works by forcing humans to do his bidding. That particular story came to a head and caused something of a moral panic surrounding the legend in 2014 when 12-year-old girls Anissa Weier and Morgan Geyser, claiming to be acting on behalf of none other than Slender Man himself, lured their friend Payton Leutner into the woods and proceeded to stab her 19 times, though fortunately not fatally.[3]

However, despite the girls' claim to have been acting on behalf of Slender Man, there is no such entity. The character was created as a "creepypasta"[4] created in 2009 by Eric Knudsen on the *Something Awful* Internet forums.[5] Because

---

3   Gabler, E. (June 2, 2014). Charges detail Waukesha pre-teens' attempt to kill classmate. *Milwaukee Journal Sentinel.* <https://archive.jsonline.com/news/crime/waukesha-police-2-12-year-old-girls-plotted-for-months-to-kill-friend-b99282655z1-261534171.html> (accessed August 9, 2024).

4   A sort of catch-all term used to describe horror-related short stories, vignettes, and images intended to be shared around the Internet, themselves following a developmental trajectory similar to those of urban legends. But in the case of these "creepypastas," the stories are usually admittedly fictional and never intended to be taken as truth.

5   Dewey, C. (June 3, 2014). The complete history of 'Slender Man,' the meme that compelled two girls to stab a friend. *The Washington Post.* <https://www.washingtonpost.com/news/the-intersect/wp/2014/06/03/the-complete-terrifying-history-of-slender-

the character is an effective horror villain, the stories took off from there and Slender Man became a viral sensation among certain segments of the Internet. As for how the two girls came to think they were acting on behalf of this fictional character, there's probably no way to be absolutely certain, but it's worthy of note that at least one of them (Geyser) has been diagnosed schizophrenic and was documented even before this incident to have conversations with other fictional characters.[6] Given that preteen girls are known to be particularly (though by no means uniquely) susceptible to at least certain forms of peer influence and social contagion[7], a plausible hypothesis is that Geyser's illness sparked the initial plan for the crime and Weier followed along for social reasons. Regardless of the actual psychological cause(s) of the crime, the Slender Man mythology grew even more prominent following this event, now even being the subject of a 2018 horror movie bearing the entity's own name.

The fact that such a wide-reaching mythology can be traced to a single internet post by Knudsen is remarkable and demonstrates the power of the kind of social transmission responsible also for the prevalence of certain urban legends and paranormal claims. But even with that history in mind, the story isn't complete. Fictional creations like Slender Man don't come out of nowhere whole cloth. Every story, no matter how creative it may be, is built upon a foundation laid by prior stories. In the case of Slender Man, there are numerous potential progenitors. Knudsen himself, in an interview with a pseudonymous blogger, identified H. P. Lovecraft, Stephen King, William S. Burroughs, horror video games such as *Resident Evil* and *Silent Hill*, Zack Parsons, "The Rake" (another

---

man-the-internet-meme-that-compelled-two-12-year-olds-to-stab-their-friend/> (accessed August 9, 2024).

6   Hale, K. (2022). *Slenderman: Online Obsession, Mental Illness, and the Violent Crime of Two Midwestern Girls*. New York: Grove Press.

7   Though social contagion theory is well-documented in certain situations, it remains controversial in others. More ordinary forms of peer pressure are more universally acknowledged. Age and sex differences across both dimensions are the subject of ongoing research, but a good case can be made that adolescents are more prone to such influences than adults and that males and females experience these pressures or contagions differently. For example:

Kravetz, L. D. (2017). *Strange Contagion: Inside the Surprising Science of Infectious Behaviors and Viral Emotions and What They Tell Us About Ourselves*. New York: Harper Wave.

Harmansson-Webb, E. B. (2014). 'With Friends Like These…'. The Social Contagion of Non-Suicidal Self-Injury Amongst Adolescent Females. Doctoral Dissertation, University of Otago.

Allison, S., Warin, M. & Bastiampillai, T. (2013). Anorexia nervosa and social contagion: Clinical implications. *Australian and New Zealand Journal of Psychiatry, 48*(2): 116-120.

Dishion, T. J. & Tipsord, J. M. (2011). Peer contagion in child and adolescent social and emotional development. *Annual Review of Psychology, 62*: 189-214.

Altikulac, S., Bos, M. G. N., Foulkes, L., Crone, E. A., & van Hoorn, J. (2019). Age and Gender Effects in Sensitivity to Social Rewards in Adolescents and Young Adults. *Frontiers in Behavioral Neuroscience, 19.*

"creepypasta" popularized on the *Something Awful* forum), paranormal stories of "shadow people" and the Mothman as inspirations.[8] Though not cited by Knudsen himself (and without any direct evidence that he even read the book), Slender Man also bears striking resemblance to the titular villain in Mary SanGiovanni's 2007 novel, *The Hollower*.[9]

Slender Man aside, how are we to conceptualize the development of urban legends (including paranormal lore) more generally? The aforementioned game of telephone goes a good job of sketching the process in intuitive terms, but fortunately we can dig a bit deeper into the matter. In our own research on both paranormal stories in particular and urban legends and folklore in general, we've identified several characteristics that we think make a particular story compelling (and therefore successful).

Plausibility. Though an argument can be made that the best stories of this genre are decidedly implausible due either to their supernatural character or to the extraordinary circumstances depicted, a successful story must contain some seed of plausibility. It might describe a rare event, but it must be believable enough that someone could think, out of all the people on the planet, maybe it really did happen to one of them. Sometimes this corresponds, even if the urban legend is fictitious, with a kernel of truth at the heart of the story that has been embellished over time and retellings.

Emotionality. Often the most successful urban legends are those that derive their compelling character from a strong emotional reaction. Some of the most effective emotions in the development of these kinds of tales include: fear/horror, disgust, greed, humor, schadenfreude/justice, or mystery. Often the emotional content is magnified by presenting the story as a sort of cautionary tale or morality play (don't commit this kind of error or that terrible thing from the story might happen to you).

Variability and anonymity. In the case of paranormal lore, this criterion may not apply to the same extent because many paranormal stories are focused on specific people and places. However, in the transmission of urban legends in general, anonymity often protects the story from the prying of skeptical researchers and variability of circumstances can make the story seem more applicable to a wide variety of readers or listeners.

Communicability. If a story is going to propagate, it must be transmissible in both content and format. Stories that are too long, too complex, too boring, etc., are less likely to be passed along. To be effective, the story should be capable of being remembered and retold without too much effort. Often this means the most effective tales will be relatively short, incredibly vivid, and feature the other factors we've discussed. The Internet has also increased the communicability of many stories by creating formats that can be shared, either verbatim or with modifications, with little more effort than the click of a mouse.

In order to more fully grasp how these tales "go viral," in the parlance of

---

8   "Slenderman235" (2011). Interview with Victor Surge, creator of Slender Man. <https://slenderman235.wordpress.com/2011/10/20/interview-with-victor-surge-creator-of-slender-man/> (accessed April 28, 2024).

9   SanGiovanni, M. (2007). *The Hollower*. New York: Leisure Books.

our times, a good subject of study would be meme theory or memetics. Though this is an entire social science in and of itself and therefore a complete treatment of the subject would require multiple volumes rather than a few mere paragraphs in this essay, we'll at least sketch the basics.

Memetics has on some level been a part of our social science for a long time, but informally so. The word "meme" was coined by biologist Richard Dawkins in his classic book *The Selfish Gene* to describe "the new replicator, a noun that conveys the idea of a unit of cultural transmission, or a unit of imitation."[10] By that, Dawkins meant to suggest that, like genes are biological replicators and serve as the unit of genetic evolution, memes can be thought of as cultural replicators which serve as the unit of cultural evolution. They're called replicators because they're capable of reproduction, and they're units of evolution because they contain information which is largely preserved across reproductions but with occasional alterations that can be subjected to selective forces. The "Internet" meaning of the word meme, denoting those widely shared and often humorous images, are a good example, but the scientific meaning of the word is not limited merely to those on-screen blips. Rather, any discrete unit of cultural information can be considered a meme—a tune, a stanza of poetry (or an entire poem), an idea, an artwork or motif, and so forth.

Dawkins' idea, which has been carried forth and developed to much greater depth by subsequent researchers, is that there is a kind of survival of the fittest in terms of cultural evolution just as there is in biological evolution. Fittest, in this sense as in biology, does not necessarily mean the most truthful or morally best. It means "best at replicating into the next generation." In that way, we can see how the telephone game of urban legends is really a game of memetics. The successful meme—a high-quality urban legend—is not necessarily one that's true, but is one that is most likely to be repeated many times, carrying itself forward into subsequent generations.

Characteristics making for a successful meme are precisely those characteristics already outlined above, but perhaps with one additional one: the ability to work together with other memes. Genes, in biology, do not exist in isolation, but in the context of larger genomes. A gene that might confer great benefit on one particular organism might, if implanted into another organism's genome, cause catastrophe. Memes, similarly, can only be considered successful if they're able to work within the broader cultural context. Further, within that context, individual memes needn't exist in isolation, but may accumulate into collective "memeplexes," analogous to genomes, which are more suited to transmission together than they would be separately.

To provide a paranormal example, the meme of "orbs" (see Volume 1, Chapter 24 for a more detailed discussion) might be meaningless and rather unsuccessful in isolation. What on Earth is an "orb" anyway, absent context? But within the memeplex of "alleged ghostly phenomena," the orb meme has proved quite successful. Hardly a day goes by in which we don't come across some new orb photography, either in our email inboxes or in the media.

---

10   Dawkins, R. (1976, 2006). *The Selfish Gene: 30th Anniversary Edition*. New York: Oxford, UK & New York.

From a Jungian perspective, one can even argue that there's something perhaps more fundamental than the meme. Or perhaps that there's a special class of meme so successful that it deserves its own designation. Those would be the archetypes. In Jungian psychology, archetypes are universal models of personality or behavior so fundamental that they do not vary from person to person but are embedded within what Jung called the collective unconscious. That's not to suggest a supernatural mechanism linking minds together unconsciously, as some erroneously interpret Jung's term, but rather that these archetypes are so ancient and fundamental to the human mind that we hold them in common and have inherited them from our most ancient of ancestors; that is, they're innate within our unconscious minds.[11] The extent to which these archetypes are truly universal may be debatable. Certainly some ideas are fundamental to all humans, others are fundamental to broad cultural groups (such as the Western world), others still are fundamental to nations or states, and so on down the hierarchy of specificity all the way to the individual. We needn't settle the debate of which entities may or may not be archetypal here, but rather only to recognize the concept. For what it's worth when it comes to potential archetypes in paranormal lore, British Museum curator Irving Finkel seems to suggest that a variety of supernatural claims, especially those involving ghosts, may be fundamental components in human culture.[12]

In that light, the most successful of memes, and here we return to urban legends and paranormal claims, are those that imitate archetypal material because in that way they are particularly resonant to the human mind—and not just an individual's human mind, but *the* human mind.

We can return to Josephina's Restaurant for a potential example. In case you haven't read (or don't remember) our previous volume, the ghost story in brief is that the property was once a speakeasy whose owner hired an assassin to murder his daughter's suitor. Unfortunately, the hitman was overly successful and killed the daughter as well. The owner's wife, grieving the loss of her daughter, is the spirit alleged to haunt the building. Of course, as we discussed in some detail in our previous book, that story was entirely made up for publicity purposes. But rather than merely dismiss it as "just a made up story," it's worth considering *why* that particular story, as opposed to so many others, resonated so deeply with so many people that it's still repeated even after it's been revealed to be false.

An argument can be made, we think, that the story touches on matter that, if not archetypal, at the very least belongs to a class of highly successful memeplex. Similar stories of fathers (or their hitmen) targeting unwanted suitors only to accidentally kill their daughters in the process fill the pages of fiction and paranormal lore alike. One particular example we gave in connection to the Josephina's story was the legend of the White Lady of Kinsale, a remarkably similar tale that took place as long ago and far away as 17th Century Ireland. Something about these stories is powerful to us because they cut to the very heart of the human

---

11   Jung, C. G. (1959, 1980). *The Archetypes and the Collective Unconscious.* New York: Princeton University Press.

12   Finkel, I. (2021. *The First Ghosts: Most Ancient of Legacies.* London, England: Hodder & Stoughton.

experience, dealing with issues of romantic and familial love, murder and grief, and so forth.

Tracing the origin of urban legends and paranormal claims, then, is not only a matter of trying to find corroborating historical documents to determine the validity of a particular story (though that's absolutely an important part of the process), but also of trying to understand which elements of the story have so resonated with people that it's been deemed worthy of remembrance and repetition. Sometimes, it's even a matter of separating the memes within a particular story to determine which ones may actually have occurred in reality and which were grafted on in the retelling, either intentionally or simply through an evolutionary process, to make the story conform to some archetypal standard.

1

Even Scared the Ghosts Away: Denver Firefighters Museum

Located in the heart of Denver at 1326 Tremont Place is a remarkable structure and the location of one of our most recent investigations. Formerly Denver's Fire Station No. 1, this amazing historic property is now home to the Denver Firefighters Museum, housing a carefully curated collection of firefighting apparatus and local history from over 150 years of firefighting history. Indeed the building and its collections are so remarkable, they don't even need a ghost story for us to recommend you pay them a visit.

But indeed they do have ghost stories aplenty, including tales of murder victims, former firefighters, children who died in fires, and more.

Figure 1.1. The Denver Firefighters Museum. Photo: Bryan Bonner.

The History

The Denver Firefighters Museum began its life as Denver's Fire Station No. 1 and remained in active service until the 1970s, at which time it became a museum, housing artifacts not only of Station No. 1 but of the entire Denver Fire Department and the firefighting profession in general. In that sense, the history of the Museum is also a history of firefighting itself, particularly as practiced in the Mile High City.

Firefighting has not always been the organized profession we recognize today. Early in our nation's history, extinguishing fires—like many public service activities—was a community effort. No less a figure than Benjamin Franklin himself organized what amounted to the first American fire company in 1736, decades before American independence. He was motivated by a 1730 fire that started on a ship, spread to the wharf, and proceeded to consume all the warehouses and multiple residences. As you'll see throughout numerous case files in this series, American history is full of such catastrophic fires, due to the prevalence of wood construction in Colonial and early American history, extending through the mid-1800s. In Franklin's case, the great man first did what he often did: he turned to his pen and wrote an essay in the *Pennsylvania Gazette* expounding upon the need for society to organize crews of men to prevent and extinguish fires. His vision came to fruition on December 7, 1736 when he co-founded the Union Fire Company, affectionately known as the "Bucket Brigade," staffed largely by members of Franklin's "Junto," the informal intellectual and social society he founded. The initial American fire brigade was staffed by 26 volunteers, each equipped with six water buckets and two linen bags to rescue property, but the organization quickly grew and similar companies popped up across the continent.[1]

In the case of Colorado, obviously such volunteer fire crews were not organized in the time of Benjamin Franklin. After all, our state was not founded until long after Franklin's death (Franklin died in 1790, the Colorado territory was established in 1861, and Colorado achieved statehood in 1876). Nevertheless, the founding of Denver's first fire company follows a similar history.

Though an informal volunteer fire company had been organized in Denver as early as 1862, no formal firefighting service existed in the city during the mid-1800s. On April 19, 1863, the need for such a service became abundantly clear when a stove was overturned[2] in a saloon known as the Cherokee House, located between Market and Larimer streets, starting a fire. Fueled by strong winds, this fire would destroy more than 115 buildings (which were primarily constructed of wood) representing approximately 40% of the city's business district, making it the largest fire in the city's history. It's now remembered as the "Great Fire of 1863." Witnesses described seeing flames, propelled by the winds, leaping across streets. Volunteer firefighters did eventually respond and were able to control the

---

1   Benjamin Franklin Historical Society (n.d.). Union Fire Company. *Benjamin Franklin Historical Society.* <http://www.benjamin-franklin-history.org/union-fire-company/> (accessed April 29, 2024).

2   Some blamed arson rather than an accidental overturning of a stove, but there's no real evidence to suggest the fire was started intentionally.

conflagration by about 5:00 the next morning, but it immediately became clear that this fire could and should have been prevented. Had the city been equipped with a professional firefighting operation (a so-called "hook and ladder" company) capable of responding in a more timely and organized manner, the fire likely could have been contained to the Cherokee House saloon as it initially grew slowly. Lacking that organization, however, the fire eventually grew out of control before anyone could respond.[3]

The city responded both by transitioning away from wood construction[4] and by creating the first fire department. The original Fire Station No. 1 was built in 1866 at 1534 Lawrence Street. This location was eventually closed to allow for a historic monument honoring the pioneers who settled in Denver to be built at its location. In 1881, the City hired its first professional (paid) fire fighters.

In 1909, the building on Tremont Place was constructed, following a design by architect Glen W. Huntington. The building is a two-story brick structure covering approximately 11,000 square feet and was built for the approximate price of $20,000 (which we estimate to be equivalent to some $685,000 in 2024). At the time of its construction, the firefighters still used horse-drawn apparatus, but the building was built with the intention of eventually transitioning to mechanized fire engines. Though the Denver Fire Department (DFD) purchased two "motorized triple combination apparatus" to begin this transitional period, the building was also outfitted with horse stables and a hayloft. Indeed, this building bore witness to some of the most amazing changes throughout the modernization of firefighting over the decades.

By 1924, the transition from horse drawn fire wagons to motorized fire engines was complete, and a period of renovation began for the building, including installation of an updated electrical system. The second-story hayloft was replaced with a kitchen and locker room, which still exist as part of the museum today.

In 1932, the building was again modified. New footings were added to the basement to accommodate the heavier machinery now being used, the wood floor was replaced with a new concrete one, and equipment lockers and garage doors were added, allowing the firemen to park their massive fire engines inside the station, both making maintenance easier and allowing for quicker responses to fire alarms.

---

3   Kennedy, A. (n.d.). Great Fire of 1863. *History Colorado: Colorado Encyclopedia.*
<https://coloradoencyclopedia.org/article/great-fire-1863> (accessed April 29, 2024).

4   In a tragic irony, the city had instituted a ban on wood construction in 1862, heeding warnings published in local newspapers that the city was becoming a "tinderbox." However, the ban was repealed that November, less than half a year before the Great Fire. Lest anyone become too melancholic over the irony, though, it likely wouldn't have made any difference. The ban applied to new constructions, not to the hundreds of businesses already standing. The Great Fire would have happened whether or not the wood construction ban was repealed that November. The greater failure in this instance should be seen as the lack of a formal firefighting company. However, the wood construction ban was reinstated following the Great Fire as *one part* of the city's larger fire prevention campaign.

Following this period of renovation, though firefighting as a profession continued to evolve, the building remained largely unchanged until it was decommissioned following 66 years of service in 1975 and marked for demolition. The new Denver Fire Station No. 1 is now located in a substantially larger property (to accommodate the growing city) not too far away near Colfax and Speer.

However, Myrle Wise, then chief of the DFD, intervened and secured protection for the building as a historic landmark. It was placed on the National Register of Historic Places in 1979[5] and, following some renovations to make it accessible to the public, began its new life as a museum on May 27, 1980.

From 1982 to 1993, a restaurant operated on the second floor called the "Old Number One Fire House Restaurant," serving sandwiches, salads, and desserts. Though no longer operating as a restaurant, the kitchen remains functional to this day (and is where we had our pizza during breaks on our own investigation so as not to risk bringing food into areas housing priceless historic artifacts).

In 1993, "professional museum trained staff" were hired to run the museum.

On May 5, 2012, one of the museum's fire trucks, a 1953 Seagrave model designated Engine No. 4 and affectionately known as Old Pumper 4 caught fire while parked outside the museum. Estimates for the repairs were in the range of $35,000 to $40,000. Donations funded the repair process and Old Pumper 4 was back in working order on January 25, 2015. Except for this period of restoration, Old Pumper 4 has served double duty from 1998 to the present as a public relations rig and as a funeral rig, having been modified to carry caskets during funeral services. Indeed, on the day of our investigation, Old Pumper 4 had just returned to the Museum after serving in the annual Fallen Fire Fighter Memorial Service in Colorado Springs.

Figure 1.2. Old Pumper 4. Photo: Robert Lewis.

---

5   Hall, B. (1979). Fire Station No. 1: National Register of Historic Places Inventory – Nomination Form. [Register No. 84129589]. *National Register of Historic Places.* <https://catalog.archives.gov/id/84129589> (accessed August 9, 2024).

The museum today includes numerous exhibits in addition to Old Pumper 4. The first floor is dominated by displays of fire engines and firefighting apparatus from various points in history. It also features a theatre-like education center and a gift shop near the front desk.

Upstairs, in addition to the kitchen and restrooms (the women's room having been converted from the old chief's office; during the building's early history, no women's room was necessary as firefighting was an exclusively male occupation), the museum features displays of firefighters' lockers containing period-appropriate items from various times throughout the department's history as well as dormitories, captains' quarters, a "family room" where firefighters might have enjoyed a meal with visiting family members, and a haunting 9/11 memorial which includes a piece of an I-beam recovered from the wreckage of the World Trade Center and given to the DFD in thanks for their support during the aftermath of that horrible day.

Figure 1.3. Memorial to 9/11 firefighters. Photo: Robert Lewis.

A particularly haunting item included among the second floor displays is a blanket recovered from a house fire. The cloth has been stained a dark gray by smoke from the fire except for a lighter central section clearly displaying the silhouette of the person who'd been lying on the blanket at the time of the fire.

The basement, not open to the public, contains staff offices and numerous storage rooms including the archives of the Denver Fire Department, with records dating throughout the department's entire history. The basement is also home to countless pieces of firefighting apparatus, uniforms, and protective equipment that have been retired from active service.

Though the entire building is home to fascinating history, some of it quite haunting, the rows and rows of archives and apparatus in the basement are particularly thought-provoking. We were thrilled during our own investigation not only to see the museum's public displays but to spend some time with these lesser-known unsung pieces of firefighting history. As the museum director told us during our tour, "everything down here has seen death."

Figure 1.4. Smoke-stained blanket. Photo: Robert Lewis.

Paranormal Claims

There are numerous ghost stories associated with the Denver Firefighters Museum. The earliest known story we've been able to track down comes from a director of the museum who served in that capacity in the early 2000s. Early in her tenure as director, she had just completed a tour for some school children. She went to the breakroom located next to the men's restroom and heard the sound of water in the men's room constantly turning on and off.[6] Assuming some children were running amok in the Museum, she went to investigate but found no one in the room.

A similar story was reported by a local ghost hunting group who preceded us to the Museum. The director also told a similar story related to another occasion in which she heard sounds of conversation rather than water running from the men's room.

One ghost hunting group claimed that a camera near the "parts room"

---

6   We're not sure why monkeying with water faucets seems to be a favorite pastime of the spirits, but this is a claim we've heard on numerous cases. Perhaps the ghosts of people who died before the invention of indoor plumbing are fascinated by this technology, or maybe they're just looking for a way to get our attention. In at least some cases, faulty plumbing or mischievous children may also be to blame.

(which we have designated "Caleb's Closet" for reasons outlined below) would not operate correctly and would slide down its mount and not lock in place while in the area.

The same former director has also told stories of several other locations in the building. She claims to be sensitive to paranormal phenomena and believes the spirits were attached to her and therefore most active when she was present. Soon after she began experiencing these sounds, as well as a variety of what might be described as poltergeist phenomena (copiers ejecting paper even when turned off, bells ringing, etc.), she contacted an unspecified paranormal investigation or ghost hunting team who investigated the building with the aid of an alleged psychic.[7]

This psychic identified three particular ghosts or paranormal entities of note, but did not conclude that these are necessarily the *only* spirits haunting the museum.

The first of these spirits is known as Thomas, Tom, or sometimes Old Tom. He's been accused of being a bit of a prankster causing people to hear disembodied voices and footsteps, moving objects, causing issues with office equipment and lights, and he might or might not have been responsible for the water turning on and off. According to the story, he was the last of the horse handlers from the station's early history and caused this activity because he was looking for his horses.

According to our interview with the aforementioned former director, there was one incident in which new insulation was being hung in window wells, but every time staff returned to that location, it had been torn down. On advice of a psychic (it wasn't clear if this was the same psychic who travelled with the ghost hunters, but we think it was) who claimed Old Tom was looking for his horses, a picture of a horse was hung nearby, and this seems to have calmed the paranormal activity.

The second spirit is known as Caleb. According to that initial psychic, this was a man who had either Down's Syndrome or some kind of speech impediment. During life, he was horribly bullied and was murdered and buried during the late 1890s at the site where the fire station was eventually built.

Unlike Tom, who presents as more of a prankster kind of spirit, Caleb is said to be an angrier ghost. According to the psychic investigator, his body is still beneath a particular closet in the basement (which we have designated Caleb's Closet), and his activity is the reason the concrete floor never sets properly. Indeed, the concrete floor in that closet is substantially "bumpier" and less even than everywhere else in the building. Additionally, Caleb supposedly doesn't like when anyone leaves this closet door open. The former director reports feeling uneasy whenever entering that particular closet. However, following work with psychics and other paranormal investigators, his spirit is supposed to have substantially calmed down.

---

7   Many ghost hunting groups travel with self-professed psychics (or "sensitives"— essentially the same thing). While we're open-minded to the possibility of psychic phenomena, we can't consider their reports dispositive unless and until their own psychic abilities are sufficiently documented and demonstrated.

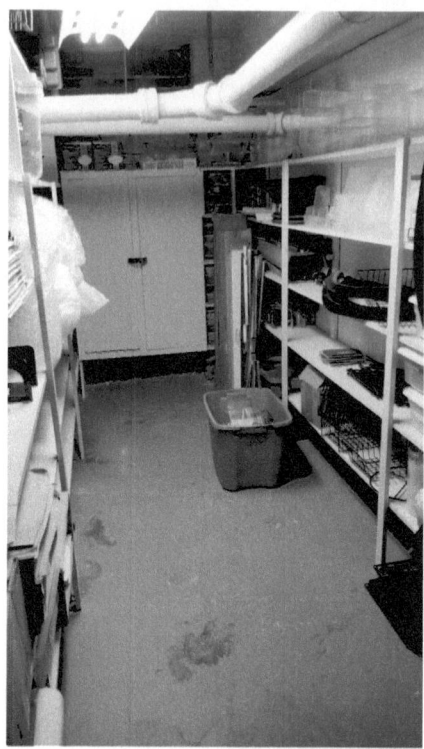

Fig 1.5. Caleb's closet has an uneven floor. Photo: Robert Lewis.

Finally, the same psychic claimed that the spirits of two children, one male and one female, of unspecified age, who were killed in a fire chose to follow the firefighters back to the building when it was still an active fire station. These spirits are said to be active following any school field trip or other event in which large groups of children are present in the museum. These children are supposed to manifest primarily on the building's second floor and are another possible culprit behind the turning on and off of the water in the restrooms.

Members of the Spirit Paranormal Investigations (SpiritPI) team clam they captured a "spectral image" or "partial manifestation" of a ghost in the basement using an infrared camera. The ghost is described as a skeletal figure with outstretched arms that has the outline of leg and hipbones and a spine but covered in a gauzy white blur suggestive of clothing. This individual is said to be reaching toward a bookshelf.[8]

In 2021, Rocky Mountain PBS aired a report following all-female paranormal investigation team XX Paranormal Communications during their investiga-

---

8   Though likely protected by Fair Use doctrine, we've opted not to reproduce the photograph here for copyright reasons, but it can be viewed in:
King, K. V. (2010). Family Fun, 10/23/10. *The Denver Post.* <https://www.denverpost.com/2010/10/22/family-fun-102310/> (accessed April 29, 2024).

tion of the Museum.[9] They reported several allegedly paranormal phenomena including a flashlight turning on and off, boots from one of the displays moving, voices coming through the Spirit Box (a device marketed as a means to record Electronic Voice Phenomena, or spirit voices), and dowsing rods moving.

During an interview with Code 3 Paranormal, a Museum staff member claimed that she was once pushed down the rear stairs by a spirit.[10] The women's restroom upstairs used to be part of the chief's office, and the story is that the spirits of the old fire chiefs don't like women wandering into what was once their office to use the restroom.[11]

While guiding our tour of the facility, the current museum director told us of the only weird or potentially paranormal experience he has personally witnessed during his time at the Museum. He said he's normally very careful to ensure all the doors are closed and locked when not in use to protect the artifacts. One day after doing some work in the storage room, he returned to find the door open. On the storage shelf is a whisky decanter which had somehow come uncorked, with liquid spilled on the floor around it.

With regard to Old Pumper 4, which now serves as a funeral rig, we expected there might be a number of ghost stories. The aforementioned former museum director was unfamiliar with any related to this engine specifically, though a variety of prior ghost hunting expeditions at the Museum have reported anomalous phenomena surrounding the engine.

During our interview with said director, she told us that following her return to work at the museum after staffing changes in the late 2010s, she had not experienced any more of the alleged paranormal phenomena. While she doesn't know the reason for this recent development, she joked that a former staff member no longer associated with the Museum (whose identity is known to us but shall be omitted from this book for reasons of privacy) may have been so horrible she scared even the ghosts away. We think that's one of the best explanations for a lack of new paranormal phenomena we've ever heard and have decided to make it, at least for our purposes, an official part of the Museum lore.

Our Investigation

Though we had interest in the Denver Firefighters Museum after seeing the aforementioned PBS special following XX Paranormal Communications' investigation, we officially opened our investigation after the Museum reached out to us regarding hosting one of their events in October of 2022, in anticipation of Halloween. We agreed to do so on the condition that we could first conduct a private investigation of our own so that we could present a firsthand account of our

---

9   Colorado Voices (2021). A night at the museum with paranormal investigators. *PBS*. <https://www.pbs.org/video/xx-2ywdwb/> (accessed April 29, 2024).

10 McGarry, E. (2018). Code 3 Paranormal – Denver Firefighters Museum – Insane Spirit Box! [YouTube Video]. < https://www.youtube.com/watch?v=OkDAxQq8EVc> (accessed April 29, 2024).

11 This is understandable. We wouldn't want someone doing their business in our offices either.

findings. They readily agreed and arrangements were made for Rocky Mountain Paranormal to conduct a one-night investigation with a small crew consisting of three of our members plus one officer of the Museum (who would participate only as an observer) on the evening of September 17, 2022 (extending into the morning of September 18).

Our team arrived on site at approximately 6:30 pm. Following a tour of the premises, during which we photographically documented the entirety of the basement, first floor, and second floor, we began setting up our monitoring equipment. Keeping one's mind focused on work when surrounded by so many historic artifacts is difficult, but we're nothing if not professional.

On the second floor, we placed a collection of toys (a plush dinosaur and two balls) in the hallway near the central staircase). These objects were meant to be potential playthings for any spirits of children as well as control objects whose location and orientation we could consistently monitor for any unexplained movement.[12] Additionally, we placed two cameras: one in the locker room with a view of the exterior of the women's restroom and one with a view of the main hallway along with the top of the stairs and the collection of toys. This second camera also captured a view of the men's restrooms. A microphone was placed at the top of the stairs.

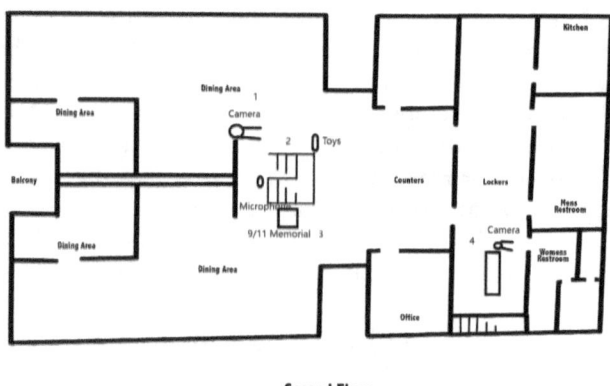

Second Floor

Figure 1.6. Second floor map. Image: RMP archives.

On the first floor, we placed one camera to monitor the rear stairway. A microphone was also placed in the same area.

Finally, in the basement we placed one camera just outside Caleb's Closet with a view through the open door. A microphone was placed nearby.

---

12   As you may recall if you read Volume 1, control objects are extensively photographed at the beginning and end of an investigation as well as consistently monitored on video so we can determine whether any movement occurs.

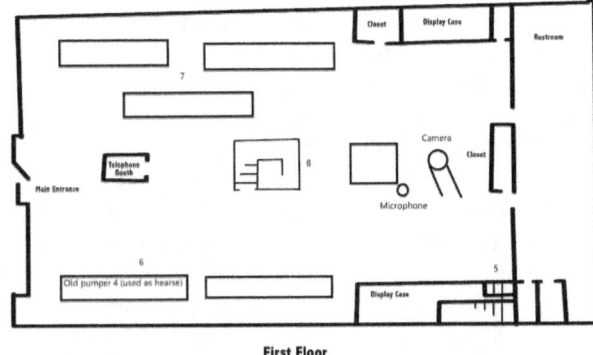

First Floor

Figure 1.7. First floor map. Image: RMP archives.

The figures we've provided show the locations of our monitoring equipment. It is important to note that these maps are not drawn to scale and do not show every display within the Museum. Additionally, the figures show the locations, market with the numerals 1 through 11, at which we took EMF readings throughout the evening of the investigation. We established our base of operations and set up our monitoring equipment on the first floor near the central staircase.

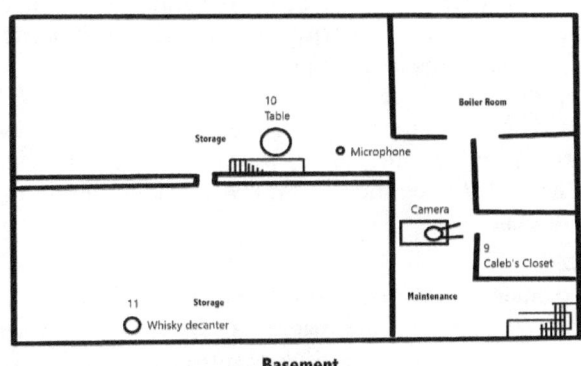

Basement

Figure 1.8. Basement map. Image: RMP archives.

The locations for EMF readings (which are also marked on the preceding maps) were as follows:
- Location 1: Second floor near the camera at the top of the stairs,
- Location 2: Second floor near the toys/control objects,
- Location 3: Second floor near the 9/11 memorial display,
- Location 4: Second floor locker room outside of the women's restroom,

- Location 5: First floor at the bottom of the rear stairway,
- Location 6: First floor near Old Pumper 4,
- Location 7: First floor near a hand-drawn pumper,
- Location 8: [Omitted from EMF readings as explained below],
- Location 9: Basement in Caleb's Closet,
- Location 10: Basement near a table by the stairwell, and
- Location 11: Basement storage room near the whisky decanter.

Though no EMF readings were taken at Location 8, that location, situated on the first floor at the top of the central basement stairway, was used for seismograph readings at the same time as the EMF readings were being taken.

EMF readings were taken several times throughout the evening, at 9:30 pm, 11:00 pm, 12:00 am, and 1:00 am. There are two kinds of EMF detectors. One, the AC meter, detects artificial electromagnetic fields such as those produced by electronic equipment, lights, etc. The other is a natural meter, which is designed to detect minute fluctuations in the Earth's magnetic field.[13] During our investigation, we measured both classes of electromagnetic field.

Readings on the AC EMF meter ranged from 0.5 µT (microTesla) to 4.1 µT, consistent with the use of electric devices and powerlines throughout the building. The 4.1 µT reading, recorded at approximately 9:30 pm near the 9/11 memorial display at Location 3 was an outlier. However, it is not elevated enough above the other readings to be considered anomalous. EMF levels in this range are not high enough to pose any health risk or cause potential hallucinations.[14] Interestingly, we noticed that the AC EMF levels consistently fluctuated slightly during our readings (and we do mean *slightly;* we're talking about tiny fluctuations occurring in real time while we took our measurements—imagine the needle "vibrating"). This is consistent with the meter detecting the transmission of data, most likely from the Museum's security and surveillance systems. We don't find any of the AC EMF readings to be anomalous.

Readings on the natural (or DC) EMF meter ranged from 0.0 µT to 1.1 µT, with the majority of measurements consisting of 0.0 µT readings. Results at each location were steady across the course of the evening, and these readings are consistent with natural fluctuations. We don't consider any of the natural EMF readings to be anomalous.

---

13   Specialized training should be obtained before using either kind of meter on a paranormal investigation. However the "natural" or DC meter, because it is calibrated to detect such minute fluctuations, is particularly susceptible to false readings in this context. Such a device should not be handheld because the operator's movements will cause EMF fluctuations. Indeed, we've even seen these meters detect the movements of an individual casually walking down a hall at the opposite end of a building.

14   As you may recall if you read Volume 1, electromagnetic fields are ubiquitous and even high levels are generally not a health risk except in the most extraordinary of circumstances (despite what you may read online). However, under some conditions and for some people with specific predilections, high EMF levels can cause hallucinations or feelings of a supernatural presence. A noteworthy case in which EMF was implicated as the cause of a client's vivid hallucinations is tentatively slated for publication in Volume 3.

While some other groups have reported EMF anomalies during their research, we were unable to replicate these findings. It's possible either that other teams may have achieved erroneous results through incorrect handling of equipment (a common problem particularly with the natural EMF meters) or that other teams were able to document phenomena that were merely absent during our investigation.

Seismograph readings taken at approximately 9:30 pm, 11:00 pm, 12:00 am, and 1:00 am at Location 8 at the top of the central stairway between the basement and first floor revealed no measurable seismic activity, even while members of our team were walking the building on the first and second floors. While the wooden floors of the second floor and stairway are surely susceptible to vibrations from footsteps, the concrete floor of the first floor does not seem to vibrate naturally, with footsteps, or as a result of passing traffic or other environmental elements. Therefore, we conclude that reports of paranormal activity at the Museum are most likely not the result of misattribution of seismic activity.

During the 12:00 am walkthrough of the building and collection of EMF readings, we also swept the entire building (with a particular focus on the second floor kitchen and basement furnace room) for gas leaks. We detected no measurable traces of gas anywhere in the Museum. In addition to being welcome news from a health and safety perspective, this also suggests that claims of paranormal activity or uneasy feelings in certain rooms are unlikely to be due to gas exposure. While we cannot rule out the possibility that exposure to gases at other times may have caused some of these reported events, it seems highly unlikely.

During our time in the Museum (both at the time of the investigation and during our visits during normal business hours) we did not see, hear, or feel anything unusual. Throughout the evening of our investigation, we took hundreds of still photographs as well as hours of video and audio recordings. Often, people get excited at the idea of paranormal investigation and ask us how they can get started. We don't like to discourage people, but one word of warning we always offer is that, done properly, paranormal investigation is science. Now, science itself is (we insist) quite exciting. But its practice also includes periods of drudgery. At the conclusion of our investigation, at least one of our members (and often more than one, so as to double check) is tasked with the duty of carefully reviewing all of the video and audio footage. Imagine sitting alone in a dark room, staring at a screen with headphones on for hours and hours, carefully watching as nothing happens. Exciting stuff!

Why would we do such a thing? Aside from pure scientific duty, part of what keeps us going are those moments when it seems like we've really found something interesting. Such an occurrence happened to one of our members upon reviewing the video footage after this investigation. While watching the camera pointed down the second floor hall toward the men's room, he saw a shadow pass across the small gap at the bottom of the men's room door. He paused the video, rewound, and watched it again several times to make sure it wasn't just a figment of his imagination. It wasn't! Something really did cast a shadow beneath that door. Thrilled, he prepared to call the rest of the team and exclaim, in his best *Ghostbusters* imitation, "we got one!" But prudence won the day and he checked the team's activity log before making that phone call. Far from being evidence of the paranormal, he realized he'd just watched video of himself wandering into the men's room during a break. Yes, sometimes we even scare ourselves.

Ultimately, we did not observe anything anomalous in any of these recordings.

A note is in order regarding the audio recording, however. One thing we observed during the evening is that the building itself is fairly noisy. Sounds of traffic and voices from people walking nearby can regularly be heard from inside the building. Additionally, one of the exhibits includes a 911 dispatch scanner that intermittently plays live dispatch calls, and these messages are also audible on our recordings.

Though it did not become relevant during our own investigation, we noticed that the building, while clearly of different construction than its nearest neighbor, is attached to an apartment building on the Northeast side.

Figure 1.9. Could sound carry over from this attached building? Photo: Robert Lewis.

It's entirely possible that some sounds or disturbances from the apartment building could be heard from within the Museum, and these could potentially account for some alleged paranormal phenomena.

Because some of the spirits said to haunt the Denver Firefighters Museum include at least one pair of children who, as the story goes, followed firefighters back to the station after being killed in a fire, we brought children's toys to see if we could record any movement of these objects during the investigation.

These objects were photographed extensively both at the beginning and end of the evening to determine whether they moved throughout the evening. Additionally, they were under constant video surveillance throughout the evening. Upon reviewing the video, none of the three objects were seen to move at all. This is confirmed by the objects' identical orientation in the before and after photographs. On a side note, you'll see in the photographs that among these toys used as control objects is a plush dinosaur character. This little guy has some history of his own, as he was rescued from the old Children's Hospital during our investigation of that property before its destruction. That story is one of our favorites and will be in Volume 3 of this series.

Figure 1.10. Control objects before (left) and after. Photo: Robert Lewis.

We have no doubt that there have been instances in which children have been killed in fires, and the story that their spirits may have followed the firefighters is a compelling one. Our lack of evidence from this investigation does not debunk this claim, but we have found no evidence to confirm it either. Because no names or dates are attached to these alleged spirit children, we are unable to search public records or newspaper articles for any historical evidence corroborating any part of the story.

With regard to Thomas, Tom, or Old Tom, the alleged spirit of the final horse handler to work at Fire Station No. 1 prior to the complete transition to mechanized fire engines, we have found little evidence to confirm or disconfirm the story. We did not observe any anomalous phenomena during the investigation that might be attributed to this spirit, but that does not necessarily discredit the story.

We have attempted to locate employment records, historic photographs, or archived duty rosters that might at least confirm whether the last horse handler was named Thomas or Tom. So far, we have not yet been able to find any records to confirm or disconfirm this claim. Therefore, this remains an open line of inquiry for us. If such evidence ever does come to light following publication of this book, we'll either release an updated edition or publish an addendum.[15]

One of the most tangible stories has to do with the alleged spirit of Caleb, who is said to haunt the basement "parts room," which we have designated "Caleb's Closet" on our maps and in our records.

Supposedly part of the reason Caleb is a restless spirit is that he's buried under this basement closet and doesn't like people walking over his grave. One of the manifestations of his activity is supposed to be that the concrete floor never sets properly. Indeed, during our investigation, we were able to note that the concrete floor in this closet is substantially more uneven than anywhere else in the Museum.

---

15   Same goes for every investigation. We're always glad to update our findings in light of new evidence.

It's important to be careful of terminology here. The concrete is "set" in the sense that it's fully cured and solid. However, we are unable to determine the reason why this particular floor is less even than those in the rest of the Museum.

In order to conclusively confirm or disconfirm the existence of an unmarked grave beneath this closet, it would be necessary to either dig up the floor or to use ground penetrating radar, neither of which we have done. However, it is worthy of note that according to the stories, Caleb was a man with either Down's Syndrome or a speech impediment who was murdered and buried on this site in the late 1800s. With regard to that portion of the story, we do have several thoughts.

First, it is plausible that someone from that era may have had Down's Syndrome, either diagnosed or undiagnosed. Though its chromosomal origin was not discovered until 1959, the syndrome itself was first described by John Langdon Down in 1862. It's also plausible that someone with either this condition or a severe speech impediment may have been bullied, perhaps fatally so. However, we have as of yet been unable to locate any police records or newspaper articles documenting the alleged crime. Records keeping in the 1800s was not what it is today, so that doesn't *disprove* the story, but neither does it offer any real historical support.

It does seem likely, though, that there would have been some mention of the case in the historical record if a body were unearthed during the building's construction. And it seems incredibly unlikely that a criminal would have buried a body so deep that it would be resting below even the building's basement. Traditional burials are six feet down (well within what is now the Museum's basement), and it seems probable that a rushed burial to hide a murder would have been even shallower. The claim that the reason Caleb haunts this closet because it's directly above his body's resting place is therefore highly suspect. Furthermore, no evidence of a body buried under the basement was discovered during the 1932 renovations to replace the floor with concrete.

That's not to say that Caleb never existed, nor even to say that his spirit doesn't haunt the Museum. But we find it extraordinarily unlikely that he haunts the Museum specifically *because* his grave is still beneath it. Such a claim seems unlikely on its face and the building's construction argues heavily against it.

One addendum to this story has to do with prior ghost hunting teams' reports of equipment failure in or near Caleb's Closet. Since we were not present for these investigations, we cannot comment on potential causes for their equipment failures.[16] However, we did experience a (rare) minor equipment failure of our own. We had initially planned to place two tablet computers with cameras in or around Caleb's Closet to record remotely. Unfortunately, both tablets failed (one seemed to undergo a factory reset and the other would not activate its camera). We do not allege that these were paranormal phenomena. The tablets were outdated and had been subject to various abuses over the years, and thus seemed ripe for failure. But in the interest of complete documentation, we note the equipment failure for the record, since we are unaccustomed to equipment

---

16   Some ghost hunting teams seem like technological wizards. But in at least some other cases, we're pretty sure that technical ineptitude is a more parsimonious explanation than anything supernatural.

failures during our investigations. That the failure occurred in a location with a reputation for causing technical glitches is at least a bit suggestive.

Some notes seem to be in order regarding the findings of other teams that preceded us to the location. Because we were not present for any other teams' investigations, we can neither confirm nor deny most of their findings. However, we offer a few notes in the interest of presenting a complete report.

The "spectral image" from Spirit Paranormal Investigations (mentioned earlier in this report) is undoubtedly one of the eerier photographs we've seen. However, given its low resolution and because we weren't present during its capture to observe either the conditions under which it was taken or the equipment with which it was taken, its evidentiary value remains inconclusive. Close examination of the photograph reveals a plausible natural explanation. The "skeletal" image to the photograph's right side appears to be a blurred image of a person walking through the room. Additionally, the "object" to the left of the skeletal image appears to be an image of one of the investigators. Close examination reveals a human face possibly wearing a hat and glasses, which we believe is one of the investigators in the photo's background.

We did not record any anomalous audio or Electronic Voice Phenomena (EVP) during our investigation. Other groups have. As with the other matters, we cannot refute the paranormal explanation but do caution that auditory pareidolia (pattern-seeking in random audio) is a known psychological phenomenon and must be accounted for when evaluating the evidentiary value of any allegedly paranormal recordings.

The American Association of Paranormal Investigators (AAPI) used a device called the Xcam SLS during their investigation which seemed to record a humanoid anomaly invisible to the naked eye.[17] The device in question is a motion tracking device modeled on the same technology as the Xbox Kinect product and is specifically designed to detect the shape and motion of human bodies. We have previously raised serious questions concerning the use of this device in paranormal investigations because a) the technology underlying the device was never intended for this purpose, b) use of the device requires a well-lit environment (which, while consistent with our own methods is typically inconsistent with other paranormal investigation groups' methods), and c) the device is designed specifically to interpret ambiguous stimuli (such as might occur in a low-light setting) as a human shape.

We recommend you watch the video from the AAPI investigation (obtained from the YouTube video linked in the footnote) and then compare their results with our own experimental image using the same class of technology to observe one of your authors' old family photos in a low-light setting.

Notice that in the AAPI video, the software has interpreted ambiguous images from the background as a human shape in the same way that the software interpreted ambiguous images to complete human shapes by adding arms and legs to our experimental image. Meanwhile, the AAPI team member in the fore-

17   Saunderson, J. (2015). Denver Firefighter Museum Xcam SLS 12.13.2014 9:43:54 PM [YouTube video]. <https://www.youtube.com/watch?v=Nhj3hOJPOiE> (accessed April 29, 2024).

ground of the image and some of the people in our own photo were not detected by the software at all. All of which is to say, we don't know for sure whether SLS technology can detect ghosts, but we *do* know that it can easily generate false positives and that its results are therefore questionable.

Figure 1.11. SLS image showing a false positive for ghosts. Photo: Robert Lewis.

Other phenomena recorded by other teams such as flashlights turning off and on or objects moving are beyond our ability to definitively comment. While it is possible to obtain such results through a loose connection between the flashlight and the battery in the former case and by accidentally moving objects without noticing in the latter case, we weren't present to determine whether these anomalies were or were not recorded under properly controlled circumstances, so we have to take them as delightful ghost stories of uncertain evidentiary value.

This case represents something of a rarity in the Rocky Mountain Paranormal Research Society's case files in that we believe our background research has successfully identified the origin of the ghost stories. Usually the origins of these stories is long since lost to history, but the haunted reputation of the Denver Firefighters Museum appears to have begun relatively recently.

Both through our scouring of records (both print and online) and through our phone interview with a former museum director, we believe that director is the first person to report concrete and specific ghost stories in the Museum. In our interview, she indicated that prior to beginning her work at the Museum, she was unaware of any ghost stories at the location. However, early in her time there, she began to experience a wide variety of allegedly paranormal phenomena.

Also in our phone interview, she confirmed that the identities of the alleged spirits Tom and Caleb as well as the unidentified children were initially reported by a psychic who accompanied the first ghost hunting team to investigate the property. Unfortunately, she could not remember the identities of the individuals involved so we were unable to reach out to them for comment.

Following these revelations, other ghost hunting groups have investigated the property, with varying degrees of success, and tend to report variations on the same stories we believe to have originate with the aforementioned psychic.

The director has indicated that after her return to the Museum in the late 2010s, she no longer experiences these paranormal phenomena. The reason for the change remains unknown to her and to us. Her lack of recent experiences is consistent, however, with our inability to document the kinds of anomalous phenomena reported by herself and the other ghost hunting groups in the past.

Interestingly, the XX Paranormal Communications investigation took place in late 2021, during the period in which the former director claims to have stopped experiencing paranormal phenomena and not long before our own investigation which yielded a similar lack of positive results. However, the XX Paranormal Communications team does claim to have experienced some paranormal events, which may call into question the allegation that supposed paranormal activity at the Museum has ceased in recent years. We would be interested to know whether any members of the XX Paranormal Communications team continue to observe allegedly paranormal phenomena during any future investigations in which they might take part.

Because the identities of the supposed spirits at the Denver Firefighters Museum were initially provided by a self-professed psychic and psychic phenomena themselves remain unproven, one's belief or disbelief in the specific identities of the spirits is likely to correspond with one's belief or disbelief in psychics. None of the findings of our investigation to date prove or disprove any of the identifications made by this psychic. However, we do note that the timeframe of Caleb's alleged death and burial and the timeframe of the basement's construction calls into question at least one part of the story.

Spending time among the history and artifacts housed at the Denver Firefighters Museum was a truly amazing experience for our team. Ghost stories notwithstanding, some of the artifacts on display within the Museum are truly haunting in the most profound sense.

With regard to ghostly activity, our investigation was unable to locate any definitive evidence. None of the alleged paranormal phenomena that have happened over the years manifested during our time at the Museum. Since we did not observe any phenomena, we are unable to provide any explanations, whether natural or supernatural, for the various claims of paranormal activity over the years.

Given that we have traced many of the ghost stories to a particular individual's early experiences at the museum and that she also reports cessation of paranormal activity in recent years, one possible explanation for our lack of results is that our investigation missed the window of opportunity for paranormal phenomena. Perhaps the ghosts, if they were ever at the Museum, have indeed gone elsewhere. On the other hand, perhaps entirely naturalistic phenomena could account for the various reported events. Since we weren't present to witness these happenings, we can only speculate. Though our background research calls into question certain aspects of the story of Caleb's spirit, we have been unable to confirm or refute the majority of historical claims related to alleged spirits haunting the Museum.

Until further evidence presents itself, we leave the case of the alleged paranormal activity at the Denver Firefighters Museum open. We hope to investigate further at a future date. For now, we leave the Museum much as we found it: a truly magnificent museum, a treasure trove of history, and home to some wonderful ghost stories of, as is so often the case, uncertain veracity.

References & Further Reading

Benjamin Franklin Historical Society (n.d.). Union Fire Company. Benjamin Franklin Historical Society. <http://www.benjamin-franklin-history.org/union-fire-company/>

Brick, C. (2018). *Haunted America: Ghosts & Legends of Colorado's Front Range*. Charleston, SC: The History Press.

Colorado Voices (2021). A night at the museum with paranormal investigators. PBS. <https://www.pbs.org/video/xx-2ywdwb/>

Denver Firefighters Museum (n.d.). "Mission and History." Denver Firefighters Museum. Accessed 22 September, 2022 from <https://denverfirefightersmuseum.org/mission-and-history>

Garcez, A. R. (2008). Colorado Ghost Stories. Placitas, NM: Red Rabbit Press.

Hall, B. (1979). Fire Station No. 1: National Register of Historic Places Inventory – Nomination Form. [Register No. 84129589]. *National Register of Historic Places*. <https://catalog.archives.gov/id/84129589>

King, K. V. (2010). Family Fun, 10/23/10. The Denver Post. <https://www.denverpost.com/2010/10/22/family-fun-102310/>

Kreck, D. (2000). *Denver in Flames: Forging a New Mile High City*. Golden, CO: Fulcrum Publishing.

Leggett, A. A., & Leggett, J. A. (2016). *No. 1: The History and Hauntings of the Denver Firefighters Museum*. [self-published].

McGarry, E. (2018). Code 3 Paranormal – Denver Firefighters Museum – Insane Spirit Box! [YouTube Video]. <https://www.youtube.com/watch?v=OkDAxQq8EVc>

Saunderson, J. (2015). Denver Firefighter Museum Xcam SLS 12.13.2014 9:43:54 PM [YouTube video]. <https://www.youtube.com/watch?v=Nhj3hOJPOiE>

2
Ghost Lights: Silver Cliff Cemetery

Paranormal claims tend to be passed on by word of mouth. Only in rare circumstances can paranormal enthusiasts point to papers or articles written in reputable non-paranormal publications to support their claims. This is such a case, though the initial write-up—in no less a source than the *National Geographic* magazine—is often misinterpreted.

The tale takes place in Silver Cliff, CO—one of the smallest towns to which our investigative program has carried us, located a little more than 50 miles west of Pueblo—where the local cemetery, located at 1435 Mill Street, is said to feature a ghostly light show. In fact, probably more people have visited this particular cemetery for the ghostly activity over the years than have gone to pay their respects to the local deceased.

Figure 2.1. Silver Cliff Cemetery. Photo: Bryan Bonner.

The History

Silver Cliff is a small town, even by small town standards. According to the 2020 Census, it boasted a population of only 609 people, actually an increase over prior years. In light of that fact, it's hard to imagine that it was once the third largest "city" (if one dares use the word to describe the little towns that marked the state's early history) in Colorado. But in its heyday, it was a bustling metropolis of more than 5,000 people, many of them miners and their families, and the seat of Custer County.

The town began to grow in the late 1870s (and was formally incorporated as a town in 1879) to accommodate the miners working the Silver Cliff Mine. For the record, the town was named for the mine, not the other way around. Miners working other mines in the area also settled in Silver Cliff.

Unfortunately, not everything went great for this little mining community. Ownership of the Silver Cliff Mine passed through numerous hands over the years, many of them East Coast mine speculators rather than local interests, and their financial machinations and mismanagement prevented the mine from ever being as profitable as the then-deepest mine in Colorado should have been.[1] Likely these financial failures are what kept the town from ever becoming large. Indeed, by the 1890 Census, the town's population had dropped to only 546 people, more than an 89% reduction from its population of 5,040 reported on the 1880 Census just a decade earlier. Its population has remained more or less consistent in the decades since, never dropping below 100 people and only in the most recent census exceeding 600.

But our story is not of the town itself. Rather, we're here to talk about Silver Cliff Cemetery, established just outside the town proper in the early 1880s to accommodate the local dead. It's a single cemetery but divided into two sections, approximately half each for the local Catholics and Protestants, the latter of which is informally known as the Cross of the Assumption Cemetery. It's still active to this day and maintained by the town itself. As of this writing, the land is only about 40% occupied, even after all these years, and plots can be purchased for as little as $500 for locals or $1500 for non-residents of Custer County.[2] Even that low price represents an increase over what it was just a few years ago. In 2016, one of our members couldn't resist the temptation to own a piece of such a famously (allegedly) haunted cemetery and, after facing delays due to groundskeeping issues, snowstorms, and an administrator's hospitalization following a traffic accident, purchased a plot for only $100.

Given that almost everyone seems to have fled the town after the largest mine in the area failed to produce profit, it should be unsurprising that our history is necessarily brief. Towns this small don't always have a whole lot of documented history about which we can write.[3] But one particular incident is worthy

---

1   Plazak, D. (2006). *A Hole in the Ground with a Liar at the Top: Fraud and Deceit in the Golden Age of American Mining.* Salt Lake City, UT: University of Utah Press.

2   Silver Cliff Cemetery (2024). Silver Cliff Cemetery. *Town of Silver Cliff.* <https://www.silvercliffco.com/copy-of-silver-cliff-museum> (accessed May 1, 2024).

3   That's not to say they have no history or that it's uninteresting. We suspect many of

of note. At around 7:00 p.m. on the evening of Friday, November 13, 1885, disaster struck at the Bull-Domingo Mine, one of the numerous mines worked by Silver Cliff residents. While ten men worked deep in the mine, an explosion rocked the ground and entombed all ten miners. It's thought, based on reports of the conditions within the mine provided by a few miners who escaped the tragedy thanks to the pure good fortune that they'd been late to work that day, that this accident resulted from some hundred pounds of "giant powder" being stored atop one of the boilers to warm and one of the primers coming into contact with the boiler, sparking the entire stock of explosives. Mine superintendent H. W. Foss was charged with criminal negligence in connection with the accident.[4]

The ten killed miners were, in alphabetical order by surname:
- B. Baptista (married with family),
- Simon Baptista,
- Napoleon Degrosslier (married with family),
- Elmer Heister,
- John Lobby,
- Conn. Nourse,
- William Patten (married with family),
- George Smith,
- William Strong (married with family), and
- H. Westfall.

A monument to the accident and the lives lost is now a permanent feature of the Silver Cliff Cemetery.

Figure 2.2. Monument to victims of the Bull-Domingo Mine disaster. Photo: Bryan Bonner.

the strangest and most fascinating episodes in human history have passed, unfortunately without documentation, in isolated cabins, small towns, and other places the press tend not to frequent.

4   Biddeford Daily Journal (November 23, 1885). The Silver Cliff Disaster. *Biddeford Daily Journal*, Maine.

Johnson, A. J. (ed). (November 19, 1885). The Bull-Domingo Horror: Ten Miners Die by the Cross Carelessness and Cheap John Management of the Mine. *The Sierra Journal.*

Paranormal Claims

Contrary to what you might expect given what you've just read, the paranormal claims surrounding Silver Cliff Cemetery don't necessarily center on the spirits of the miners lost in the Bull-Domingo Mine explosion, nor those of any particular individual or individuals buried in the cemetery. Rather, the paranormal claims all center on a particular phenomenon known as the "ghost lights" or sometimes the "dancing lights."

According to legend, the tale of the Silver Cliff Cemetery ghost lights began some time in the early 1880s when three miners were walking home from a party and opted to take a shortcut through the cemetery. While that sounds to us like a wonderful set-up for a horror story, this case didn't end with bloody catastrophe but rather with a remarkable and potentially paranormal experience. As the miners made their way past the grave markers, strange ghostly lights, balls of blue and white color, began to flit about the gravestones, treating the witnesses to a sort of unearthly dance of luminescence. Of course they immediately rushed to town and told their story, and the legend was born.[5] Also unsurprisingly, particularly if you've read the essay which opened this section, the story hasn't always been consistent. The identities of these three miners are not known, and different write-ups of the story differ in almost every detail except for the presence of those ghostly lights dancing in the graveyard.

Often, even if the paranormal manifestation in question begins its life anonymously, people eventually tie it to a particular historical event or character. These ghost lights are unusual in that no one seems to definitively ascribe them to any particular ghost or entity, but rather enjoy them for what they are. That doesn't mean people don't speculate, though. Many people we've spoken to over the years believe they're the ghosts of at least some of the individuals who found their final resting place in the cemetery. A few have ventured to think they could be connected to the Bull-Domingo disaster, but even those who've suggested that possibility seem to do so only speculatively. One writer suggested rather than ghosts, they could be related to faeries.[6] Surprisingly, no one we're aware of has yet connected them with UFOs.

Whatever their source or nature, they've become a tourist destination for ghost hunters and paranormal enthusiasts. Numerous independent witnesses have described their experience of seeing the lights over the years, many of them quite credible. Many people have even remarked that on one particular night, the entire town of Silver Cliff agreed to turn off their lights to determine whether their own lights could be causing the phenomenon, but the ghost lights kept dancing in the cemetery unabated.

---

5   Scoles, S. (2022). What's Really Behind the Ghost Lights of Colorado's Silver Cliff Cemetery? *Atlas Obscura*. <https://www.atlasobscura.com/articles/silver-cliff-ghost-lights> (accessed May 1, 2024).

6   Quinn, P. (2005). Ghost Lights of the Silver Cliff, Colorado Cemetery. *Legends of America*. <https://www.legendsofamerica.com/co-ghostlights/> (accessed May 1, 2024).

Our Investigation

A lot of ghost stories and other paranormal claims are fairly recent, even if they deal with old spirits. That this particular story is claimed to go all the way back to the 1880s caught our attention, so in addition to our on-site investigation of the cemetery itself, a substantial portion of our investigative work in this case has been an attempt to trace the history of the lore surrounding it.

We begin with the story with which we closed the previous section: that one night the entire town shut off their lights and yet the ghost lights continued. It's an often-repeated story, but no one ever seems to be able to identify the night on which it allegedly took place so that we'd have a fair shake at finding out whether or not it really happened. However, we've never let a little bit of difficulty stand in our way, and we found what we believe to be the source most of those retelling the legend are relying on. The earliest mention we've been able to find of this mass shut-off of lights came from a quote by Custer County Judge August Menzel in the August 20, 1967 issue of *The New York Times*: "One night, everybody in Silver Cliff and Westcliffe shut off their lights at the same time. This included the street lights, but that did not affect the grave lights. They were still there, and a lot of us saw them."[7]

We have to admit, when we first heard the story, we were incredibly skeptical. Ever since we found that article, we're not quite sure what to make of it. For one thing, we have not found any record documenting the actual shutting off of the lights, and it seems like the sort of thing that ought to be documented in public record somewhere. Even the idea that the municipality could be persuaded to turn off all the lights for ghost hunting purposes, while it gives us hope for some of the future projects we have in mind, seems extraordinary. On the other hand, the tale was repeated by a presumably-respectable county judge in a major national newspaper, so take that for whatever it's worth and feel free to assess the story's credibility by whatever metric you please.

Now we turn our attention to the provenance of the legend itself. Many believers who recount the story mention that it's a rarity in that it was written up in the *National Geographic* magazine, which does at least superficially seem to lend some credibility to the tale—that publication does not, after all, have a reputation for sloppy reporting. And on the other hand, some skeptics have even suggested that the *National Geographic* write-up of the tale is the first citation of the legend.[8] Indeed, the article does exist. It was written by Edward Linehan and consists mostly of a travelogue about the State of Colorado, but the very end of the article contains a single column of text describing the author's own experience in witnessing the lights.[9]

---

7   Little, W. T. (August 20, 1969). Hunting Ghosts in a Ghost Town Out West. *The New York Times*.

8   Stollznow, K. (2013). The "Dancing Lights" of Silver Cliff Cemetery. *James Randi Educational Foundation*. <https://archive.randi.org/site/index.php/swift-blog/2005-the-dancing-lights-of-silver-cliff-cemetery.html> (accessed May 1, 2024).

9   Linehan, E. (1969). The Rockies' Pot of Gold: Colorado. *National Geographic, 136*(2): 157-201.

A few important takeaways from that article bear mentioning. First, the author does acknowledge having witnessed the lights, but also acknowledges that, though he doesn't know the solution, a natural explanation will likely someday be found. Reflection of lights from the town (one possible solution many skeptics have proposed over the years) is considered but dismissed as an unlikely but not impossible solution. Most importantly, the story is not presented as a recent development. It's acknowledged as longstanding local lore, demonstrating that this article was not, in fact, the story's point of origin, though it does seem to be the point at which the story achieved national prominence.

If you pay careful attention to the dates of the citations in this chapter's footnotes, you may have already noticed that the *National Geographic* and *New York Times* articles were both published in 1969. In fact, they were published in the very same month of August that year, suggesting that this was the point at which people across the nation (and perhaps to a lesser extent internationally) discovered the story of the ghost lights.

But not only do these articles both acknowledge a longstanding legend, they also don't represent the first citation of the story in print. The earliest one we've been able to find so far was a 1956 article in the *Wet Mountain Tribune,* a newspaper focused specifically on news of local interest.[10] We can't say for certain that this article is *the* first publication of the ghost lights story, but it's the earliest one we've been able to find so far.

Could it be that this article represents the earliest report of the legend, or could it be that the story, as some versions have it, originates as long ago as the 1880s and has been passed on by word of mouth for the decades before it started seeing print? As is too often the case, there's no definitive answer, but the text of the *Tribune* article does provide a clue: "It seems that one evening early last week, some of the younger set of the community were taking a nocturnal ride, which carried them past the cemetery. As they passed the cemetery, screams of terror came from the feminine occupants of the car as they beheld numerous eerie lights dancing about among the tombstones…The following day, tales of the ghostly lights were spreading rapidly through the community."[11]

To us, that doesn't sound like an article describing longstanding folklore. It sounds like a news piece telling locals of a phenomenon they're unlikely to be familiar with. If so, that places the origin of the ghost lights in 1956 rather than the 1880s.

So just what are these ghost lights? We hoped we could find some good photos or videos of them as a place to begin our investigation. Alas, we came up empty. There are lots of photos of the cemetery itself, but precious few that even allege to be of the ghost lights. And of that latter group, the only ones we found were what we'd consider stereotypical "orb" photos which were produced by reflecting light (most likely a camera's flash) off pieces of dust, pollen, or other particulate matter in the air (see Volume 1, Chapter 24 for a detailed account of

---

10   Wet Mountain Tribune (April 13, 1956). Silver Cliff Cemetery Ghosts Carrying Own Lights At Night: Curious Custerites Flock To Burial Ground To View Unusual Reflections. *Wet Mountain Tribune, 72*(40).

11   *ibid.*

how these photographs are produced). Dust particles, however, while they may explain the orb photographs, do *not* explain what people claim to see in person when they visit the cemetery (orbs occur only on camera and not to naked eye).

This brings us to our own investigation. One chilly night[12] we set up camp in the cemetery to see what we might witness and what we might be able to photograph. On the latter point, we were unable to capture anything of terrible interest. But on the former point, we do have some ideas.

Throughout the course of the evening, we noticed that we often saw light reflected off the tombstones, both from the town and, most prominently, from the occasional car passing on a nearby road. On that particular evening, there were even more reflective surfaces than just the tombstones themselves as someone had decorated many of the graves with blue reflectors. Whenever light from any source hit those surfaces, it put on quite a show that, we think, could easily be misinterpreted by someone who didn't know what they were looking at.

And indeed, reflected light is the most common skeptical explanation offered for the ghost lights. Even that original 1956 article considers this explanation (combined with the tendency of youngsters to try to scare the bejeezus out of each other) the most likely explanation. Such an explanation is not without its objections, though.

First, one might argue that if the story truly dates from as long ago as the 1880s, electric lights could not account for the phenomenon because they simply didn't exist then. That is a false statement. Electric lighting was first invented in the *early* 1800s. However, we do acknowledge that, especially in a small town, there probably wouldn't have been *enough* electric light to reliably produce a ghostly lightshow. This seems like a moot point, though, in light of our contention that the 1956 article marks the beginning of this story. By that time, electric lights were in wide usage, as were automobiles with headlamps.

More difficult to overcome are the objections that lights from the town can't account for the ghost lights because either the town is too far away or the ghost lights occur even when the town's lights are off. The story that the entire town experimented with turning their lights off one evening and still saw the ghost lights, if true, would seem to mean the reflection argument does not actually account for the phenomenon. We simply haven't been able to determine whether that story is true or not, though.

Even the *National Geographic* article anticipated this argument, pointing out that the town *seemed* too distant for reflections to account for what the author witnessed and alleging that the ghost lights occur even when a fog separates the cemetery from the town.

Further, Silver Cliff and neighboring Westcliffe are International Dark Sky Association (IDA) recognized "dark sky communities."[13] This means the towns have replaced streetlamps with models designed to reduce light pollution and

---

12   Though at least it wasn't a blizzardy night as often seems to be the case during our investigations for some reason.

13   Healy, J. (2016). Colorado Towns Work to Preserve a Diminishing Resource: Darkness. *The New York Times.* <https://www.nytimes.com/2016/08/13/us/colorado-dark-sky-project-stars-perseid.html> (accessed May 1, 2024).

passed ordinances requiring outdoor lights to point downward. As keen amateur astronomers often distraught at the amount of light pollution, particularly around larger cities, we love that some people are working to keep the lights dim enough to render the night sky visible. But it also calls into question the "reflections from town" hypothesis regarding the ghost lights. If the towns are reducing their light, it seems like this would stop the ghost lights if indeed they're caused by these reflections.

There are a few possibilities. We do note that the IDA recognition and dark sky initiatives are a relatively recent development, so it's possible reflections from the town from earlier years could have accounted for some of the lights. That's not our favorite hypothesis, though. It seems more likely to us that it's not reflections from the towns themselves that have caused the ghost lights. If that were the case, the lights ought to have been a constant occurrence (at least prior to the dark sky measures), which simply wasn't the case. Rather, they were intermittent. To us that suggests that, if reflections are the culprit, it's more likely reflections from passing car headlights than from the towns themselves.

Indeed, during our investigation, we didn't see any "dancing lights" consistently throughout the evening but we did take note that the graveyard lit up with reflections whenever a car passed by, particularly if the driver had activated the high beams.

Another possibility, in light of the area's noted darkness (if you'll forgive the pun), is something known as the "prisoner's cinema." We previously described this phenomenon in greater detail in our discussion of the Cave of the Winds (Volume 1, Chapter 6), but in brief, it's a phenomenon in which an individual subjected to a long period of low light or dark will begin to hallucinate a "cinema" of dancing lights. This results from the eyes and brain, not ordinarily conditioned to such darkness, trying to make sense of their surroundings. It's entirely possible that many of the people reporting the ghost lights were actually experiencing prisoner's cinema.

Not all accounts can be dismissed as hallucinatory, though. There have been reports of multiple witnesses seeing the same lights. Two or more people might simultaneously experience prisoner's cinema, but it's unlikely they would experience the same hallucinations at the same time. Thus, while we're all but absolutely convinced that at least some of the ghost light reports could have been due to this phenomenon, we're also all but absolutely certain that not all of them were.

Where does that leave us? Well, we're not entirely certain. We didn't witness anything that looked paranormal to us during our own investigation, but we did see some reflected lights in the cemetery. At the end of the day, we have no way to prove a negative or to say the ghost lights don't exist, but we can say that we didn't see them and that we're pretty sure at least *some* of the reports (if not all of them) can be accounted for by natural means.

That said, don't let us turn you off from a visit to this cemetery. Whether it's haunted or not, it's a beautiful location and dark enough to be pleasant. If you get lucky, maybe we'll turn out to be wrong and the ghost lights will put on a show for you. But even if not, at least you'll get a view of the night sky far superior to what those of us doomed to live in the cities or suburbs are cursed to live with.

References & Further Reading

Biddeford Daily Journal (November 23, 1885). The Silver Cliff Disaster. *Biddeford Daily Journal*, Maine.

Healy, J. (2016). Colorado Towns Work to Preserve a Diminishing Resource: Darkness. *The New York Times*. <https://www.nytimes.com/2016/08/13/us/colorado-dark-sky-project-stars-perseid.html>

Johnson, A. J. (ed). (November 19, 1885). The Bull-Domingo Horror: Ten Miners Die by the Cross Carelessness and Cheap John Management of the Mine. *The Sierra Journal*.

Linehan, E. (1969). The Rockies' Pot of Gold: Colorado. *National Geographic, 136*(2): 157-201.

Little, W. T. (August 20, 1969). Hunting Ghosts in a Ghost Town Out West. *The New York Times*.

Plazak, D. (2006). *A Hole in the Ground with a Liar at the Top: Fraud and Deceit in the Golden Age of American Mining*. Salt Lake City, UT: University of Utah Press.

Quinn, P. (2005). Ghost Lights of the Silver Cliff, Colorado Cemetery. *Legends of America*. <https://www.legendsofamerica.com/co-ghostlights/>

Scoles, S. (2022). What's Really Behind the Ghost Lights of Colorado's Silver Cliff Cemetery? *Atlas Obscura*. <https://www.atlasobscura.com/articles/silver-cliff-ghost-lights>

Silver Cliff Cemetery (2024). Silver Cliff Cemetery. *Town of Silver Cliff*. <https://www.silvercliffco.com/copy-of-silver-cliff-museum>

Stollznow, K. (2013). The "Dancing Lights" of Silver Cliff Cemetery. *James Randi Educational Foundation*. <https://archive.randi.org/site/index.php/swift-blog/2005-the-dancing-lights-of-silver-cliff-cemetery.html>

Wet Mountain Tribune (April 13, 1956). Silver Cliff Cemetery Ghosts Carrying Own Lights At Night: Curious Custerites Flock To Burial Ground To View Unusual Reflections. *Wet Mountain Tribune, 72*(40).

3
Ghosts of Old Miners: The Phoenix Mine

Located at 800 Trail Creek Road in Idaho Springs, Colorado, about 30 miles west of the heart of downtown Denver (as the crow flies) is the Phoenix Gold Mine. This location, now operated as a tourist destination, offers a glimpse into Colorado's gold rush past. As with most old gold mines from the 1800s, the Phoenix Mine has seen more than its share of death. And today, it's become one of the most talked about yet relatively seldom investigated alleged haunts in the state. Permanent residents of the mine are thought by paranormal enthusiasts to include the spirits of miners who died in accidents, murder victims, and maybe even some tommyknockers.

Figure 3.1. A Phoenix Gold Mine shaft. Photo: Bryan Bonner.

The History

Anyone who knows anything about Colorado history can tell you that the state's early history is overwhelmingly the history of pioneers who moved west looking for gold. An old joke has it that Denver itself is a testament to laziness as early west-bound settlers, upon seeing the formidable Rocky Mountains, promptly said "to hell with this" and settled where they were rather than brave the fraught journey across the Continental Divide. In reality, though, it was opportunity rather than cowardice or laziness that kept the early Coloradoans in the state. As formidable as they are, those same Rocky Mountains are home to remarkably rich deposits of gold, silver, and numerous other valuable minerals.[1] Numerous fortunes have been made in these mountains. Unfortunately, mining's profitability corresponds with its dangerousness, and likely for every fortune made in this trade, another life has been lost.

Such is also the case of the Phoenix Gold Mine (which we'll shorten to simply the Phoenix Mine or the Mine for convenience throughout this chapter).

Digging up the Mine's natural history (separate from any consideration of paranormal claims) has become a bit difficult. Web searches, usually a good first step (though rarely the only step) in any historical investigation, are dominated by tourist advertisements and paranormal stories. For us, that's not a terrible thing. Paranormal stories are, after all, what we're ultimately looking for. Still, tracking down the non-paranormal part of the story took us a bit longer than is often the case. But we're nothing if not stubborn and we were able to find some interesting facts about the Mine itself.

The ore vein that would eventually become the Phoenix Mine was discovered in 1871 by a man whose last name was, fittingly enough, Miner. He eventually sold the mine to two Cornish miners who, after some rough starts, eventually struck it rich and worked the mine until the outbreak of World War I. The miners abandoned the mine and returned with their fortune to Cornwall, England. It was then purchased for the royal sum of $20 owed in back taxes by a local real estate agent who proceeded to "salt" the mine by firing a shotgun loaded with gold into the dirt. He then sold the property for the substantial profit of $5,000 to the Gunderson family, who worked the mine even deeper and, finding a deeper gold vein, struck it rich for themselves. Unfortunately, in 1943, President Franklin Delano Roosevelt, in order to support the World War II effort, shut down all the gold and silver mines, attempting to direct those resources (both in manpower and in machinery) to military purposes. Given that the same President Roosevelt had, in 1933, all but banned the possession of gold, few of the mines reopened after the war.[2]

In the years to follow, the mine passed through a few hands. One group of owners didn't continue mining operations in earnest but were able to obtain

---

1   Including, of relevance in more recent decades, the third largest uranium reserves in the United States.

2   Oakes, D. (2003). GPR Tours the Phoenix Gold Mine. *Gold Prospectors of the Rockies.* <https://www.goldprospectorsoftherockies.com/articles/phoenixmine_031025.htm> (accessed May 6, 2024).

enough gold with relatively little effort to buy a new car, at which point they sold it to a Mr. Al Simmons sometime in the 1950s who then eventually sold it to Al Mosch and family in 1972. Though Al Mosch died in 2019, his family (now in their fourth consecutive generation of Colorado mining) still own and operate the Mine which has functioned as a tourist destination since 1988, though they estimate that throughout its history the Mine produced over 100,000 troy ounces of gold.[3] We did a quick "back of the envelope" estimate of that gold's value based on current gold prices as of this writing and determined its total value to be more than $230 million in 2024.[4]

Throughout its history, the Phoenix Mine has seen its share of both positive and negative happenings. On the positive side, it has been host to at least one wedding we're aware of. And in the 1980s, the Vatican sent a delegation of some 30 bishops and one cardinal who blessed the Mine when they learned of the Mosch family's plans, in collaboration with the Missionary Sisters who'd recently visited from Italy, to raise funds for orphans of Colorado miners.

But on the darker side, it's also known that there are human remains within the mine. In one case, the remains are in the form of ashes scattered in a roped-off area near the main tunnel. These ashes belong to a visitor of the mine who died (not at the mine) of lung cancer but wished for his remains to be scattered in a place to which he felt a strong connection. However, in at least one other case, the remains belong to a miner who was buried during a mine collapse. Evidence of that collapse can still be seen to this day.

Figure 3.2. Evidence of the collapse can still be seen. Photo: Bryan Bonner.

Current mine staff and tour guides will also confirm that several fatalities have occurred at the Mine over the years. Unfortunately, that's to be expected in such a dangerous occupation. We know precious little about any of the victims of those accidents over the years, so we can't confirm how many such fatalities there

---

3   Phoenix Gold Mine (n.d.). About Us at the Phoenix. *Phoenix Gold Mine.* <https://www.phoenixgoldmine.com/about-us> (accessed May 6, 2024).

4   We're in the wrong business.

were nor the victims' identities. What we can say is that there is probably no such thing as a mine operated as long as the Phoenix Mine without numerous fatalities over the years. Even in modern times, with improved technology and safety standards, underground mining continues to be one of the most deadly civilian occupations, with a frightening 20.1 work-related fatalities per 100,000 employees, according to a 2022 report by the U. S. Bureau of Labor Statistics.[5] And in the past, the dangers were even more pronounced and included mine collapse, explosion, gas poisoning, heavy metal toxicity, machinery accidents, suffocation, and a whole host of illnesses related to the working conditions. And that doesn't even count the occasional murder committed over the mines' products.

In fact, that last point leads us directly to the transition between confirmed history and paranormal claims.

Paranormal Claims

The Phoenix Mine is home to, perhaps, more than its share of ghost stories as well as some other paranormal claims. In fact, though relatively few people have actually conducted investigations into the alleged paranormal occurrences at the Mine, it's one of those locations that pretty much all of the local paranormal enthusiasts have at least heard of in passing. Often when we give our public lectures, it's one of the places about which some of our audience members have questions or want to share an anecdote they heard. Sometimes we inadvertently prompt those questions when we discuss our various credentials because some of our members are, in fact, certified miners.[6] Others have read about our work at the Cave of the Winds (Volume 1, Chapter 6) and want to know whether we've done any other underground research. Well, indeed we have, and the Phoenix Mine is a perfect example.

Probably the most famous source of paranormal claims related to the Phoenix Mine comes from an episode of *Ghost Adventures* in which Mr. Bagans and company explored the mine.[7] The episode touches on a lot of the very same ghost stories we were told by members of the Mine staff when we were present, so that seems like a good place to start. In that episode, the paranormal enthusiasts claim to have encountered a variety of potentially paranormal occurrences

5   USBLS (2022). Civilian occupations with high fatal work injury rates. *U. S. Bureau of Labor Statistics.* <https://www.bls.gov/charts/census-of-fatal-occupational-injuries/civilian-occupations-with-high-fatal-work-injury-rates.htm> (accessed May 8, 2024).
6   We're not kidding around when we say that paranormal investigation draws upon all possible professional backgrounds. In the case of the mining certification, we didn't pursue it specifically for paranormal investigation purposes—many of our certifications come from our various day jobs—but you're crazy if you think we're above exploiting it when we want to go play in an old haunted mine.
7   Gallo, P. (producer) (2016, September 24). Colorado Gold Mine (Season 13, Episode 1). [TV series episode]. In Gallo, P. (producer). *Ghost Adventures*. MY Entertainment.

including EVP[8] and even some manifestations that appeared on video.

The most "complete" story surrounding some of these alleged ghosts in the mine was told by one of the Mine's owners and involves a double homicide committed just outside the mine by a man named Gillespie in or around the 1950s. Supposedly Gillespie was a mentally deranged individual who shot two uranium prospectors[9] in the head for no apparent reason. The story goes that Gillespie then spent the rest of his life in a mental hospital and the two victims are supposed to haunt the property. The *Ghost Adventures* episode includes an alleged spirit photograph of two ghostly individuals who may or may not be the same two individuals who were murdered.

One of the things that particularly got our attention was that the Phoenix Mine isn't home only to spirits but also, apparently, to tommyknockers. Tommyknockers originated in Cornish folklore but have become a part of mining culture worldwide. They're thought to be diminutive humanoid creatures that might be thought of as akin to goblins or gnomes, and they inhabit and work in underground mines. Often they're thought to be benevolent and trying to help keep the miners safe, though in some versions of the lore they're given either a mischievous or sometimes even malevolent character. No lore surrounding the Phoenix Mine alleges any particular characteristics for the supposed tommyknockers, but they were mentioned in the *Ghost Adventures* episode as possible inhabitants of the mine. This view is espoused by paranormal enthusiast Chuck Zukowski, who actually accompanied Rocky Mountain Paranormal on our own investigation.[10] The *Ghost Adventures* crew even went as far as to claim they captured what they believed to be a tommyknocker on their SLS camera.

We'll address those claims in detail when we discuss our investigation shortly, but it's worth pausing for a moment to enjoy a paranormal claim different from the usual "mere" ghost story. That's not to say we don't like the ghost stories. Obviously we do; these books are filled with them. But at the same time, we get particularly excited whenever a claim comes across our desk of something other than ghosts simply because, as much as we love the ghosties, it's nice to get some variety from time to time. Tommyknockers fit that bill.

---

8   That's "electronic voice phenomena." We've explored this topic in greater detail in volume 1, but in brief this is the practice of recording "noisy" audio (either white noise or random radio signals) in the hopes that spirits can manipulate the recordings to communicate through the radio or recording device.

9   Turns out, there's a lot of uranium ore in Colorado, and uranium prospecting became a popular activity during the Cold War. It's still done today, though not to anywhere near the same extent. Murders aside, uranium prospecting isn't as dangerous as it might sound. Unrefined uranium ore is, all things considered, relatively safe to handle (particularly in small enough doses) and certainly not the sort of thing from which one could make a nuclear weapon without highly sophisticated enrichment facilities that are beyond the means even of most world governments. We even have some Colorado uranium ore in our own cabinets of curiosities.

10   Zukowski, C. (2013). Phoenix Gold Mine, Ghost Investigation 2011. *UFOnut*. <https://www.ufonut.com/phoenix-gold-mine-ghost-investigation-2011/> (accessed May 8, 2024).

Numerous visitors have reported seeing or hearing a variety of manifestations ranging from visible appearances of ghostly miners to whispered voices or EVP recordings to sounds of knocking which are alternatively attributed either to ghosts or tommyknockers. Some even claim to have been touched by a ghostly presence. Over the years, several psychics have claimed to communicate with the spirits in the Mine, and plenty of people have come up with photographic anomalies ranging from the fairly impressive (like the aforementioned visible manifestation the *Ghost Adventures* people claim to have captured) to the relatively mundane (such as orbs, about which see Volume 1, Chapter 24). Other than the story of the double homicide near the mine, none of these phenomena seem to be attached to any particular story or historical character.

One additional claim that caught our attention has been reported by a number of witnesses over the years. It seems that, from time to time, a "bright light" is known to illuminate the Mine about halfway down the main tunnel.

Our Investigation

When we set out to investigate the Phoenix Mine, we were joined by paranormal enthusiast Chuck Zukowski, who has in the years since gone on to host the Travel Channel program *Alien Highway* and now runs the *UFOnut* website. The investigation was originally conceived as part of an ill-fated attempt to pitch a television series dedicated to our own brand of skeptical investigation with Zukowski chosen to present the "believers" counterpoint to some of our more skeptical members.[11] Though the television series came to naught, we did at least get to spend some time looking for paranormal phenomena in a remarkable old mine, which really is its own reward. Some bats even joined the investigation, which is always a thrilling occurrence for our zoologically-inclined members.

Though we always take our own environmental readings, and did so in this case as well, we were informed by Mine staff right at the start of the environmental conditions we'd face on the investigation. The Mine itself is more than 1000 feet deep and reaches both the Phoenix and Resurrection ore veins (the latter named when it was reached in the 1970s to commemorate the "resurrection" of the mine). Average temperatures within the mine remain relatively constant

---

11   We've begun negotiations with a variety of producers and television networks over the years. We continue to believe that our kind of investigation could make for a good television program as long as the long stretches of boring grunt work are edited out and the show is presented with theatrical or cinematic reenactments of the original claims. Unfortunately, most producers don't seem to understand this vision and prefer instead to go for the easy and cheap thrills of sensational programming. Throughout these negotiations, various producers have insisted upon a) finding positive paranormal conclusions in every episode, b) the presence of psychics on the investigative team, and c) the presence of "a cute twenty-something goth girl" on the investigative team. We have no objections to the latter, but these demands demonstrate the difference between the kind of work we do (even though we think we can make it entertaining) and the kind of thing that gets aired on paranormal television.

year-round, fluctuating between 42 and 54 degrees Fahrenheit. Below the main tunnel is a smaller level known as the "Winze," which can be reached by means of a ladder and small tunnel.

The investigation itself lasted for several hours during a single session, and was largely uneventful, consisting mostly of our usual practice of periodic measurements using EMF meters, seismometers, and other measuring devices in between long stretches of silent video and audio recording.

For his part, Zukowski also engaged in some EVP sessions in which he attempted to communicate with spirits alleged to be present. He obtained some recordings which he believes could be evidence of either spirits or tommyknockers present in the mine, but which we consider to be inconclusive. A knocking sound he recorded, while we can't prove it wasn't a tommyknocker or some other paranormal entity, could also have just been environmental sounds within the mine. He also claims to have recorded a paranormal vocalization which we find more interesting, but also inconclusive as the vocalization can only be heard following significant refinement of the original audio track.[12]

Our own measurements and recordings didn't detect any anomalies during the time we were present, so we don't have a whole lot to go on when evaluating the paranormal claims. We did, however, make several observations that can explain at least some of the various paranormal claims people have made.

With regard to the mysterious bright light that sometimes shines in the middle of the main tunnel, we were able to identify a highly probable natural cause. There is an area in the upper tunnel directly above where the light is meant to shine which contains an opening to the Mine's exterior. A sort of skylight, if you will. While direct sunlight doesn't *always* shine through this opening, there are times when the opening is in direct alignment with the Sun's position in the sky. During these times, the entire tunnel is indeed illuminated by a bright light.

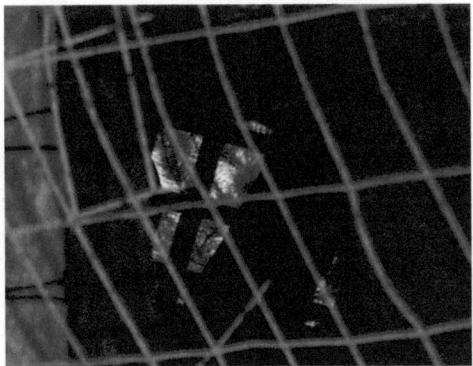

Figure 3.3. The Phoenix Mine skylight. Photo: Bryan Bonner.

Is that proof positive that there's no supernatural explanation for the lights

---

12   Zukowski, C. (2013). Phoenix Gold Mine, Ghost Investigation 2011. *UFOnut.* <https://www.ufonut.com/phoenix-gold-mine-ghost-investigation-2011/> (accessed May 8, 2024).

people have reported over the years? Of course not. We weren't present to see what those witnesses saw. However, this does seem like the most likely explanation.

Other phenomena didn't occur in our presence, but we do have some speculative ideas that provide at least potential explanations for some of them.

Throughout our investigation, we did pick up some low-frequency vibrations inside the Mine. While we didn't notice these ourselves, they were easily detected by our measuring equipment. Low frequency vibrations are known, under certain conditions, to create either hallucinations or feelings of a supernatural presence (see Volume 1, Chapter 10 for a more detailed explanation). It's likely, therefore, that at least some paranormal claims may have been caused by these vibrations, but it's equally likely that not *all* of the reports can be so explained.

The low light conditions inside the mine can also produce a "prisoners cinema" of hallucinations (see Chapter 2 as well as Volume 1, Chapter 6). It's quite likely that some of the visual manifestations could be explained by this phenomenon. It's much less likely that a full-bodied manifestation of a spirit dressed in miner's attire could be so easily written off, but those are rarer claims and ones we didn't personally witness so we can't offer too much in the way of commentary on those.

Mold could be another factor. Several locations in the mine were covered by various types of molds and fungi. Alas, we couldn't collect samples for analysis. This investigation took place before we started carrying specimen jars as a matter of routine (as we do now—live and learn), and time constraints prevented further analysis. Various molds are known to cause hallucinations, especially if one is exposed to large concentrations in a confined space. We're uncertain whether any of the ones we observed in the mine are such varieties, so we mention this only as a hypothetical explanation for some paranormal claims.

Ultimately, our on-site investigation proved fairly uneventful. That doesn't mean boring by any stretch of the imagination. Any excuse we can find to spend some time tommyknocking around in a creepy, perhaps haunted, old mine is sure to produce a good day for people like us. But in this case, there's unfortunately just not much very exciting to report.

Since the investigation, though we haven't (yet) returned for a repeat visit (something we probably ought to do in the near future if it can be arranged), but we have kept abreast of further claims. Several ghost hunting groups have produced photographic or audio anomalies we've examined but have not yet found any we consider to be convincing.

We have also attempted to follow-up on the story of the double homicide just outside the mine, but without much success. We've searched in vain for any newspaper articles or public records related to the matter, but the only citations we can find are people repeating the story anecdotally and in connection to the ghostly lore. An inquiry to the Idaho Springs Police Department in regards to this matter has, as of this writing, gone unanswered.[13] If any information comes to

---

13   A formal public records request (the local equivalent of a FOIA request) could force the matter, but lacking a date for the murder, a full name for the perpetrator, or any identifying information about the victims, we simply don't have enough information

light in the future, we'll either revise this chapter in subsequent printings and/or publish a separate addendum. But for now, since the only sources of information about the alleged murder are in the form of people telling spook stories about the Mine, we're highly skeptical that the crime took place.

Where does that leave us? Pretty much where we started. The Phoenix Gold Mine is a wonderful historic mine, a great place to spend an evening, and perhaps even home to some ghosts or tommyknockers. But as hard as we've tried, we haven't found any real evidence to demonstrate those latter points, so we have to leave this case mostly open and unsolved in our ledgers.

References & Further Reading

Gallo, P. (producer) (2016, September 24). Colorado Gold Mine (Season 13, Episode 1). [TV series episode]. In Gallo, P. (producer). *Ghost Adventures*. MY Entertainment.

Oakes, D. (2003). GPR Tours the Phoenix Gold Mine. *Gold Prospectors of the Rockies*. <https://www.goldprospectorsoftherockies.com/articles/phoenixmine_031025. htm>

Phoenix Gold Mine (n.d.). About Us at the Phoenix. *Phoenix Gold Mine*. <https://www. phoenixgoldmine.com/about-us>

USBLS (2022). Civilian occupations with high fatal work injury rates. *U. S. Bureau of Labor Statistics*.   <https://www.bls.gov/charts/census-of-fatal-occupational-injuries/ civilian-occupations-with-high-fatal-work-injury-rates.htm>

Zukowski, C. (2013). Phoenix Gold Mine, Ghost Investigation 2011. *UFOnut*. <https:// www.ufonut.com/phoenix-gold-mine-ghost-investigation-2011/>

to fill in the requisite forms.

4
A Trip to Deadwood: The Bullock Hotel

Though the bulk of our work is done in or near our home base in Colorado—after all, we're entirely self-funded so we essentially go as far as our pocketbooks will carry us—we do occasionally get the opportunity to investigate out of state. This was such a case, prompted by an invitation to spend several nights investigating the alleged hauntings of the Bullock Hotel, located at 633 Main Street in Deadwood, South Dakota (which town was itself made famous by the HBO program *Deadwood*). The Bullock hotel is the oldest hotel in the town and remains popular to this day.

Not only was this one of our rare out of state expeditions, but it's also become one of our favorite stories because our investigation presented a perfect blend of explainable phenomena and unsolved mysteries. All this despite the difficulties involved in conducting an investigation in a building that houses not only hotel guests but an active casino.

Figure 4.1. The Bullock Hotel. Photo: Bryan Bonner.

The History

As the oldest hotel in Deadwood, South Dakota, the Bullock Hotel is, unsurprisingly, home to quite a bit of interesting history. Most hotels have seen their share of both triumph and tragedy. Often the latter gets successfully swept under the rug by hotel management who are typically not too keen on the press getting wind of some of the things that happen in hotels. In the case of the Bullock, though, much of the history is extremely well-documented.

The hotel was built in 1895 by Seth Bullock and his business partner Solomon "Sol" Star for the princely sum of approximately $40,000.[1] Adjusted for inflation, that would be about $1.5 million in 2024 dollars.

Mr. Bullock himself was born in Amherstburg, Canada on July 23, 1849 to parents George (a retired British Major) and Agnes Bullock (nee Findley, originally from Scotland). At the ripe old age of sixteen years, fleeing from the strict (often corporal) discipline of his father and an unhappy family life, young Seth ran away from home to move in with his sister in Montana. According to unverified legend, his sister forced him to move back home. Whatever the reality of that situation in the Bullock family, Seth did move back home but eventually moved himself back to Helena, Montana in or around 1867.[2]

Despite, or perhaps because of, his fraught childhood, Seth Bullock was an ambitious young man. Shortly after his arrival in Montana, he unsuccessfully ran for a seat in the territorial legislature. Undeterred by his defeat, he ran again for a seat in the Territorial Senate, where he was successful and ultimately served from 1871 to 1872. During his office, he was instrumental in the formation of the Yellowstone National Park on March 1, 1872.

Such accomplishments notwithstanding, Mr. Bullock is better remembered as a lawman than as a politician. After his term in the Senate, he was elected Sheriff of Lewis and Clark County in 1873. The practice of law enforcement in those days was not the same as it is today. During the course of his duties, he confronted a man named Clel Watson who had stolen a horse. The thief responded by shooting Sheriff Seth in the arm. Mr. Watson, for his part, was promptly captured, tried, and sentenced to be hanged. According to the legend, just as Watson was placed on the gallows, an angry mob appeared and scared off the executioner. Sheriff Bullock responded by climbing up on the gallows himself and holding off the mob while pulling the executioner's lever himself.

It was during this period in his life that Mr. Bullock formed a business partnership with Mr. Star. They established the Star & Bullock Auctioneers and Commission Merchants in Helena Montana. This partnership would be long-lasting and would ultimately lead to the creation of the Bullock Hotel.

It's not clear exactly when Bullock and Star set their sights on South Dakota and Deadwood, but it's clear they were thinking about it for some time in advance of the actual move. In 1874, Bullock married his childhood sweetheart Martha

---

1   Parker, W. (1981). *Deadwood: The Golden Years*. Lincoln, NE: University of Nebraska Press.

2   Wolff, D. A. (2009). *Seth Bullock: Black Hills Lawman*. Pierre, SD: South Dakota State Historical Society Press.

Eccles, but quickly sent her and their new child back to her family home in Michigan until he could settle in his new home state.

Seth and Sol, planning to establish a hardware store, arrived in Deadwood via ox-drawn wagon and began setting up their new shop on the corner of Main and Wall streets on August 1, 1876[3], just one day before Deadwood established itself in gambling lore as the place where Wild Bill Hickok was shot in the back of the head by Jack McCall at Nuttal & Mann's Saloon No. 10.[4] Clearly, a new sheriff was needed. One Mr. Isaac Brown was appointed on August 5 by a "Miners' Court." However, Mr. Brown and his party were ambushed and killed while en route to Deadwood. The Court met again and appointed Mr. Con Stapleton, who served until March of 1877 when Governor Pennington appointed Seth Bullock as the new Sheriff of Lawrence County (of which Deadwood was and remains the county seat).

By all accounts, he was the right man for the job. It was said he wouldn't have even needed to carry a gun and that his mere presence in town was sufficient to enforce the law. His grandson would later say "he could outstare a mad cobra or a rogue elephant."[5] At one point, Sheriff Bullock's duties led to another noteworthy encounter. While bringing a horse thief known as "Crazy Steve" back to Deadwood to stand trial, he encountered the then-deputy Sheriff of Medora, North Dakota—none other than future-President Theodore Roosevelt. Seth Bullock and Teddy Roosevelt became fast friends and Seth would eventually join Roosevelt's "Rough Riders" organization and to serve as Captain of a troop in Grigby's Cowboy Regiment. However, the end of the Spanish-American war intervened and Bullock's troop never left the training camp, but Bullock carried the title of Captain for the rest of his life, and in 1905 would form a group similar to Roosevelt's in Deadwood which he called "The Cowboys." The same year, he was invited to President Roosevelt's second inauguration and attended with his Cowboys. The newly-reelected President appointed Bullock a United States Marshal for South Dakota, a post he held for nine years.

It took until 1876 for Bullock and Star to open their hardware store, but they eventually did so and it operated from a single-story building also home to a blacksmith from 1876 to 1894. In 1879, a fire in Deadwood destroyed much of the town including the majority of Bullock and Star's hardware store, leaving only the foundation and the exterior walls. Though they rebuilt, another fire in 1894 again destroyed the building's interior. This time, they decided to rebuild a bigger building and to open what would become the Bullock Hotel, featuring sixty-three

---

3   Some reports have Seth and Sol arriving on August 2, the very same day as the fateful events described below.

4   This event has etched itself permanently in the lore of card-players (and by extension, magicians). Most gamblers can even tell you the cards Hickok held when he was shot: the famous "dead man's hand" of aces and eights (though the fifth card is unknown and the subject of sometimes-spirited debate).

5   Cornelius, J. (2024). The Finest Type of Frontiersman. *Frontier Partisans.* <https://frontierpartisans.com/34762/the-finest-type-of-frontiersman/> (accessed June 15, 2024).

ten-by-ten foot rooms, each equipped with a chamber pot and with one full bath-room on each floor. They also proudly featured a steam heating system.

By this time, perhaps due to Sheriff Bullock's watchful eye, the town had begun to settle down and was deemed safe for a family. Seth thus sent for his own family and was promptly joined by his wife along with daughters Madge and Floy and son Stanley (and Stanley's own wife). For her part, Mrs. Bullock became a prominent member of the local community until health issues later in life con-fined her to her second-floor room where she would only be seen, occasionally, on her balcony.

A few other incidents from Mr. Bullock's life are worthy of note. In 1880, Seth and Sol partnered with Mr. Harris Franklin and founded the Deadwood Flouring Mill, with Seth serving as general manager. Also with Sol, he founded the S&B Ranch Company. Together they convinced the Fremont, Elkhorn & Missouri Valley Railroad to route a new line across a ranch they'd started. This resulted in the founding of the town of Belle Fourche, several miles north of Deadwood.[6] Free lots of land were offered to homesteaders looking to move into the area, resulting in the largest livestock shipping area in the country and the nomination of Belle Fourche as the new county seat of Butte County in 1894. When his longtime friend and former President Roosevelt died on January 6, 1919 (at the age of sixty years), Bullock had a monument erected in his honor on Sheep Mountain, which was dedicated on Independence Day—July 4, 1919.

Seth Bullock died shortly thereafter, on September 23, 1919 at the age of seventy years. The location of his death is uncertain and the subject of some debate. It is thought that he either died in Room 211 of the Bullock Hotel, at his ranch in Belle Fourche, or at his home in Deadwood. The location of his final resting place, however, is known. He is buried in Mount Moriah Cemetery in Deadwood, near the graves of Wild Bill Hickok and Calamity Jane. His grave, appropriately enough, faces Mount Roosevelt.

In popular culture, he was portrayed by Timothy Olyphant in the acclaimed HBO television series *Deadwood*, which aired from 2004 to 2006, and in the spin-off movie in 2019.

The hotel itself has seen plenty of action over the years, both positive and negative. A particularly noteworthy episode occurred during a local epidemic of smallpox, also known around that time as "the speckled monster." Many of our readers may be old enough to remember a time before vaccines eradicated this insidious disease.[7] Before its eradication, outbreaks occurred with frightening reg-ularity in communities around the world. Deadwood itself suffered several over the years, and during one of them, the basement of the Bullock Hotel, known now as "Seth's Cellar" became a children's smallpox ward. At this time in history, medicine was still in relative infancy and nurses were in short supply. Care of the sick children often fell to two very different groups of women: nuns and prosti-

---

6   Though this wouldn't have been known to Mr. Bullock at the time, of course, Belle Fourche is just about twenty miles south of what is now considered the geographic center of the United States following the admission of Alaska and Hawaii to the Union.
7   The last naturally-occurring diagnosis occurred in 1977 and the WHO officially listed smallpox as eradicated in 1980.

tutes.[8] In the case of the Bullock Hotel, the latter assumed the duties of seeing after the ill children. To foreshadow just a bit, this will become relevant when we discuss the ghost stories later.

Eventually the Hotel fell into disuse and disrepair. The majority of the wood structures had rotted, some rooms collapsed. Even the original roof had caved in, leaving debris scattered throughout the entire building. In 1990 a project was undertaken to restore the hotel to its former glory. It took more than ten semi-trailer trucks to remove the decades of accumulated trash and damaged building contents. However, the project was ultimately successful. While retaining as much as possible of the original structure, the Hotel was repaired and brought into the modern era. The original sixty-three rooms were reduced to the current twenty-eight, but each room is substantially larger and are now equipped with (as one would expect from a modern hotel) private bathrooms. Some even have wet bars and hot tubs. Despite the modern conveniences, the Bullock Hotel in its current incarnation retains much of the Old West charm of its original construction and now features, in addition to the aforementioned twenty-eight guest rooms, a full bar and restaurant (called Bully's after Teddy Roosevelt's nickname) and a twenty-four-hour casino.

Paranormal Claims

It's probably true that any old building has its share of ghost stories. The same is true of hotels.[9] Of course, given its placement in this book, the Bullock Hotel is no exception. Its ghost stories, in fact, are too numerous to even begin to list all of them, but as they reflect the rich and colorful history of the Hotel and the town in which it's located, we'll do our best to provide a reasonable cross-section. Some of the stories are clearly based in established history. Others fit more within the "urban legend" category of unverified and probably unverifiable tales passed down through local communities over the years.

Probably the most noteworthy (at least for us) of all the ghost stories takes place in "Seth's Cellar." This basement room now serves as a secondary bar (in addition to the one that's part of the Bully's restaurant) and event center. But as we mentioned in the previous section, Seth's Cellar once served as a smallpox ward for ill children attended by the local "working girls." Given that history, it should be no surprise that there are ghost stories connected with that room. One alleged spirit in particular is a little girl known as Sarah. She, along with her mother (whose name is not known but who is said to have been one of the ladies from the nearby Kitty's Brothel), were guests in this makeshift hospital. According to legend, Sarah watched her mother die shortly before succumbing to the disease

---

8   Say what you will about prostitution and say what you will about the Catholic Church, but prostitutes and nuns, each in their own ways, have saved the United States numerous times throughout history. Indeed, we probably owe the survival of Western civilization in no small part to these groups of women.

9   Whether they advertise it or not is another matter, but every hotel we've checked has at least one ghost story, typically of children playing with a ball in a hallway. There seems to be something almost archetypal in that kind of claim.

herself.

Over the years, Sarah's spirit has allegedly been witnessed by numerous people—particularly children. When a child is left unsupervised or wanders off alone in the Hotel, the story goes, Sarah coaxes the child into the basement with the promise of toys and a playmate. Though this sounds like the setup for the kind of horror movie we'd love to watch, the reports don't seem to indicate that Sarah's ghost has any malevolent intentions. Presumably she's just looking for someone close to her own age to play with. However, also in line with what we've seen in horror movies, some of the children also report that once they arrive in the basement, a "tall man" tells them they need to return to their parents, sometimes even helping them find their way back to their families upstairs. Several parents have even reported seeing their children wandering the halls with one hand raised into the air. When asked why they were behaving so strangely, the children have said they'd been holding the hand of the tall man.

Who is this tall man? Could it be Seth Bullock himself? That would have been our first guess, just based on the way these kinds of stories seem to go, but it turns out that seems not to be the case. Some of the children involved in these stories have shown their parents the portrait of a man they claim to be the tall man who helped them. Whose portrait did they identify? None other than former President and longtime friend of Seth Bullock, Teddy Roosevelt himself. And does President Roosevelt match the description of a "tall man," as the children have described? That's probably a matter best left to individual judgement. His height was five feet, ten inches. That puts him only slightly above the current average of five feet and nine inches (for males in the United States). However, that's certainly "tall" by children's standards and well above the average male height in the United States of about five feet and seven inches during his own lifetime.

One employee who had worked at the Hotel for well over a decade tells of a time she was preparing helium balloons (about forty-five in number) for a child's birthday party scheduled to be held in Seth's Cellar at the Hotel. While she worked, the balloons clung to the ceiling (as helium balloons ought to do). But occasionally one of them would bob up and down as if someone were pulling on the string. That seems eerie enough but she also claims that during the party, all of the balloons popped at the same instant and she has no idea how or why that could have happened.

Guest Room 303 boasts an interesting story provided by someone who was staying at the Hotel some years ago. She'd been given reservations at the Hotel as a gift from her husband just before his death and decided to keep the reservation as a way to remember him. She stayed in Room 303. One night, at around 2:00 a.m., she went to the front desk and complained that she'd been awakened by someone sitting at the foot of her bed. She described the intruder as a tall man with a hat and said that when she looked up, he "just vanished." She refused to return to the room, even to retrieve her own belongings. Hotel staff had to pack her bags and help her relocate to another room.

A similar story occurred to a different individual in (we think) a different room. She sent a letter to the Hotel describing her experiences. While waiting for her husband to come back to the room, she saw some papers "fly off of the television." She later reported that she felt what she thought was her husband sitting next to her on the bed, but when she turned around to comment about what she'd seen with the papers, she found she was alone in the room.

As you might expect, the ghost of Seth Bullock himself is thought to haunt the Hotel. Shortly before our investigation, witnesses reported seeing a full-bodied apparition of Mr. Bullock in Bully's bar and restaurant. He's also regularly thought to haunt Room 211, which had been his own room and is one of the three places reported as the possible site of his death. Scents of lilac, roses, and cigar smoke are reported throughout the Hotel without known cause. Room 211 is commonly the site of such reports of cigar smoke (as is the casino floor). Employees who've found themselves slacking off or whistling while they work have reported experiencing phenomena they attribute to an unhappy Mr. Bullock trying to encourage them to get back to work, including hearing their names called even when no one is present.

Bully's Bar has its own haunted reputation. In addition to the alleged apparition of Mr. Bullock, the bar is home to a piano that is sometimes seen and heard playing itself. Glasses and plates have been seen to "take flight" in the bar, and sometimes bar stools are said to move about on their own.

Cleaning staff have often reported that while they're going about their duties, the cart they use to carry supplies mysteriously moves to other rooms even when no other staff members are cleaning in the area. After changing rolls of toilet paper, housekeeping staff have reported returning to the restroom only moments later to discover the paper strewn about the room and the spool still on the holder. Maintenance staff have also reported that sometimes if they happened to turn on a radio while working, the radio would switch itself to a Country music station. This was reportedly Seth's favorite genre of music.

Guests and staff alike have reported that the shower in Room 208 sometimes turns on and off seemingly of its own volition.

The Bullock Suite has several claimed paranormal phenomena. The most noteworthy is a "phantom clock" said to ring on its own even though it has been broken for decades. One employee even claimed to have witnessed as the clock "jumped onto the floor by itself." This room is also home to a replica of Seth Bullock's hat, which is sometimes seen moving by itself, sometimes even floating above the entertainment center.

At the top of the main stairway is a mirror claimed to exhibit some haunting properties. This mirror was reported to be one of last pieces of original furniture still at the location. Both guests and staff have reported seeing images of Seth Bullock and sometimes Teddy Roosevelt in the mirror's reflections. It's also said that photographs of the mirror often exhibit strange "paranormal-looking" images in the glass.

We can see what people mean when they say the image in the mirror seems somehow "paranormal." To us, it looks like something out of *The Blair Witch Project*, though of course there is no actual connection between the Bullock Hotel and that movie.

Honestly, the entire Hotel seems to be home to plenty of ghost stories. But in our research, we've found that the greatest concentration of such tales seem to have originated in Rooms 205, 207, 208, 209, 211 (Seth's old room), 302, 303, 305, 314, Bully's Bar, and Seth's Cellar. Each and every one of these locations has seen reports of "standard" kinds of haunting phenomena including appliances turning themselves on and off, lights flickering or turning off and on, and the sounds of babies or children crying even when no children are present.

Figure 4.2. The haunted mirror with its arcane symbol. Photo: Bryan Bonner.

Our Investigation

Sometimes, things work out just the way we want. As we mentioned earlier, we usually don't get to work very far from our home base in Colorado, simply because we're self-funded. Turns out, if you're not being paid by a television studio, it can get really expensive to trek all over the world looking for ghosts, aliens, and all the other things that go bump in the night. This is especially true because we do not accept compensation for our investigative work (see our essay on "Ethical Considerations in Paranormal Investigation" earlier in this volume). However, we're more than happy to accept room and board if someone wants to put us up for an out of state investigation, and that's exactly what happened here.

Management of the Bullock Hotel knew all about their haunted reputation and had heard that we're good at what we do so they made us the kind of offer we'd have to be fools to turn down: they'd put us up in the Hotel and feed us for the duration of our on-site investigation. All we had to do was show up and do what we do best. We're nobody's fools so we made the proper arrangements to conduct a three-night investigation at the Bullock Hotel.

Any hotel presents some investigative challenges. There's simply too much ground to cover to be able to look at everything all at once. The Bullock Hotel presented some extra challenges on that front. First of all, we didn't want to take advantage of the Hotel's kind offer of accommodations so we traveled with a smaller crew than we might have otherwise brought had the location been closer to home, though we remained in contact with our people back at "home base" in Colorado. Second, because the Hotel and casino were still in operation during our investigation, we had to tread carefully both to avoid disturbing the normal day to day operations of the business and to ensure the other guests and gamblers didn't inadvertently contaminate any of our data.

We're always happy to rise to a challenge, though, and the Hotel's management and staff were certainly accommodating.

Our team arrived at about 2:00 in the morning (apparently we're nocturnal even in our travels). Upon arrival, we were assigned to our various rooms and set about unpacking all of our equipment. If you think business trips mean "traveling light," you're probably correct most of the time. The exception is for paranormal investigations, in which case we travel with godawful amounts of assorted scientific and surveillance equipment. After unpacking we also did an initial tour and survey of the location, started to develop a game plan, and then promptly went to bed to get some rest before beginning the investigation in earnest.

We arose at about 9:00 in the morning and reconvened. We started off by interviewing staff as well as a few willing guests staying at the Bullock to determine which locations would be best for our investigation and what kinds of phenomena we should be on the lookout for. We even took the Hotel's public Ghost Tour to make sure we had a good sense of the lore.

After a bit of discussion, we selected several locations for monitoring and established our game plan. We established our base of operations in the main casino, which is both a phrase we never expected to write and an activity, we're sure, which attracted plenty of attention from people who just wanted to play cards or slot machines.

Figure 4.3. Our base camp in the casino. Photo: RMP archives.

For the actual investigation, we selected several locations:
- Seth's Cellar, including the hallway leading from the basement bar area to the kitchen,
- The main central stairway, including the two mirrors at the top of the staircase,
- Room 303,
- Room 211,
- The back stairway leading to Bully's Bar,
- The third floor main hallway, and
- The second floor main hallway.

Of course we followed our usual protocol of establishing video and audio monitoring for much of the time while we all watched from our base of operations, punctuated by periodic walk-throughs to take EMF readings and other measurements.

Seth's Cellar got the lion's share of our monitoring equipment, including video cameras at the base of the stairs looking toward the seating area, on the bench also pointed toward the seating area, at the back of the room pointed toward the seating (this camera was later moved to view the end of the bar), overlooking the back hallway near the elevator (this camera was later moved to the main room to watch a table with control objects), at the back of the room overlooking the bar, at the back of the room overlooking the bar seating area, at the base of the stairs watching the bar, on a table next to the stairs also pointed at the bar, overlooking the kitchen area, overlooking the basement entrance area, and on the bar itself (this camera was later moved to look at the mirror behind the bar). Microphones were placed at the base of the stairs and at the back of the room.

Because the room was said to be haunted by child spirits, we placed several toys around the room as control objects. These included two green foam balls on the bar and one foam baseball on one of the chairs in the seating area as well as crayons and blank paper on the round table at the back of the seating area and on a table across from the bar.

We also used a fine dusting of cornstarch around the bar to be able to monitor any possible movements or disturbances.

In other words, we figured this room was going to be the interesting one so we made sure we would know about anything and everything that might happen in there.

But we didn't neglect the other rooms. The main stairway area was monitored by three video cameras, a microphone, a stationary EMF meter, and a foam ball on top of the allegedly haunted mirror as a control object. Room 303 was under constant surveillance by a single video camera. Room 211 got two video cameras and a foam ball on the center of the bed as a control object. Bully's Bar got one video camera on the first night overlooking the stairway and two cameras on the second and third nights overlooking the dining booths and the bar. The second and third floor hallways were each monitored by a single video camera overlooking the hallway. Finally, we had one camera mounted on a slot machine to overlook our base of operations (because it's not every day you get permission to put surveillance equipment in an active casino and we have a sense of humor).

Our investigation lasted for three nights. As is our usual protocol, we took temperature and EMF readings approximately every hour. Across all three nights, those readings were fairly straightforward and boring. Temperatures remained consistent the whole time and EMF readings were both consistent and within the range of what we expected to find in a hotel and casino.

At 1:15 in the morning on the first night of monitoring, two of our team members were in Seth's Cellar taking their hourly readings when they reported hearing a noise at the bar. They dropped what they were doing to investigate (eventually they got back to complete their readings just a few minutes late). As they approached the bar, they found the coffee machine at the end of the bar had started running. Unfortunately there was no coffee in the machine but it was running hot water as if brewing. When asked, the staff told us this machine hadn't been used in years. Dust collected around the machine corroborates that story.

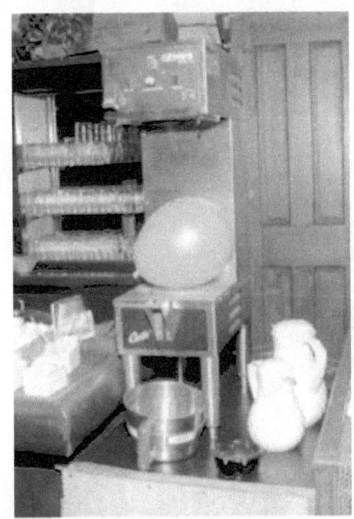

Figure 4.4. The haunted coffee machine. Photo: Bryan Bonner.

At 2:25, the camera in the hallway from the basement to the kitchen dimmed as if something had blocked the infrared sensor on the camera. However, no one and nothing appeared in the camera's image itself. Though we were unable to conclusively determine the cause of this incident, we do have to acknowledge that this is within the realm of what we could call "mundane anomalies." That is, we can't explain it, but neither does it seem outside the realm of normal behavior for electronic devices which can somethings just do funny things.

After that, we decided to rest for the remainder of the evening and packed up our equipment. Upon returning to his room (Room 307), one team member reported hearing the sound of boots on a hardwood floor just outside the bathroom. However, that area is carpeted and when he checked the hallway, no one else was present.

For the second night, we got an earlier start, motivated in part by the minor anomalies we'd experienced the night before and because we had an opportunity to engage in a longer run of investigation while the areas under surveillance were closed to the public.

Early in the evening, something happened that's haunted us ever since.

Two foam balls had been placed on the bar in Seth's Cellar as control objects, with a camera monitoring them continuously. At 8:10 p.m., one of the two balls—only one of the two, mind you, even though they were near each other—rolled approximately six inches toward the back of the bar and stopped abruptly just in front of the other ball. No one was in the bar at the time, but we were monitoring the video from our base camp in the casino and immediately went to investigate.

Figure 4.5. The haunted balls on the bar. Photo: RMP archives.

We spent an obscene amount of time rolling little green balls across a bar, trying to recreate what we'd witnessed. We couldn't find any grooves or uneven surfaces on the bar that might explain it. A level confirmed this more precisely—the bar was indeed flat and even. We tried blowing the balls across the bar, and while it was possible to do so, it required enough wind that it didn't seem like a particularly likely explanation, particularly since only one of the two balls rolled...and then stopped.

Though we couldn't recreate it immediately, we shifted our investigation toward looking into what might have caused this. We never did figure it out.

Alone, that might *sort of* haunt our thoughts. Maybe. It's weird, after all, but we've seen a lot weirder. What happened next made it so much worse and implanted the story permanently in our collective memory.

Though we couldn't recreate or explain what we'd seen, we did capture it on video. And one of the important things about the way we work is that we're always cognizant of the idea that just because *we* can't explain something, that doesn't necessarily mean it's a ghost or that *nobody* can explain it. So we immediately got online and shared the video with some of our friends and colleagues back home in Colorado to solicit their ideas regarding what might have happened. None of them had an explanation at hand. But it was the next morning that turned a curious event into an infuriating one. When we booted the computers back up to continue our work, the video was simply *gone*.

A word of explanation is in order here, because we've seen a variety of ghost hunters or paranormal teams deal with technological issues that we find pretty mundane. Computers are complicated things and it's incredibly easy to accidentally delete something, or even for a malfunction to occur and lose a file. In our case, though, it's worthy of note that several of our team members are experienced computer professionals and even they couldn't figure out what happened to the video file. Worse, when we got back home and referred the matter to data recovery professionals of our acquaintance—the kinds of people who charge an obscene fortune to recover files for governments and major corporations when loss of data simply isn't an option—they told us they couldn't even find any evidence on the computer that such a file had ever existed. If over a dozen people hadn't seen the video, it would be easy to accuse us of making things up or simply hallucinating.

We never did get the video back. The photo you see above was taken the same evening, but during our attempts at recreating the phenomenon.

However, not all was absolutely lost. At 8:21 the very same evening, another event occurred.

In addition to the little green balls on the bar, we had placed a foam baseball on one of the chairs in the seating area, and it also exhibited some strange behavior: again without anyone else in the room, it rolled *up* and off the front of the chair.

Figure 4.6. The ball that rolled off the chair. Photo: RMP *archives.*

If you look closely at the image above, you'll notice that these chairs have a sort of ergonomic design. That is, the seat is just a bit off level, and the front of the chair is elevated slightly higher than the back. As such, if some small air current or vibration disturbed the admittedly precarious position of the ball, we would expect it to roll downward toward the back of the chair. This ball did the opposite, seeming to defy gravity and move up and over the front of the chair.

Fortunately, this was also captured on video, and *these* videos did not vanish from our computers.[10] We now have those video files backed up so many ways they are probably the most redundantly reproduced and protected piece of data in our entire archives. We're not taking any chances.

At this point, we knew our third and final night of the investigation was going to be pretty heavily focused on those little balls. We weren't yet prepared to assume a paranormal explanation; we wanted to do our due diligence and see if we could come up with any other ideas. Though it means presenting this part of our investigation out of order, we'll explain what we did the following night before carrying on with the rest of the chronology.

First, we rolled and poked and prodded and blew on those balls every way we could possibly think of. No success. We used levels to make sure we weren't falling victim to some optical illusion. Sure enough, the bar was level and the chair sloped the way we thought it did. Some of our team members stomped around like crazy people to see if footsteps, perhaps combined with something like a loose floor board, could cause the balls to roll as a result of vibrations. No dice. We poured cornstarch all around the bar to see if we could detect any kind of disturbances and found nothing.

One thought that quickly occurred to us and stuck in our heads was that there might have been some kind of air current that caused these balls to roll. It would be remarkable for such a thing to roll only one of the two balls on top of the bar, but we figured stranger things have happened so that seemed at some point like the most plausible of all the natural explanations we could think of. Trouble was, none of us felt a draft and we couldn't locate any likely source of a draft. But that doesn't necessarily rule it out.

These days, we would use high-tech precision devices to measure the flow of air currents all through the room, but this occurred several years ago and we were a little lower-tech at that time (plus, we were traveling and didn't have all of our gear with us). So we came up with the next best idea we could: the next morning we sent one of our team members to a local store to buy every helium balloon we could find and tied them to all of the chairs in the bar. We then spent the rest of the night watching for any draft of air to move one or more of the balloons.[11] None of them ever moved except under the direct action of one of our team members.

These balls—these stupid little balls—remain one of our most intriguing

---

10   Ask us at one of our public appearances and we'll gladly play you the video.

11   We must have made quite the sight. Imagine a group of grown adults staring at a bunch of balloons like their lives depended on it. Yeah, we get ourselves into some weird situations.

and frustrating mysteries to this day.

Returning to night two to finish up our narrative (though the most interesting part is now out of the way), after the movements of the balls, one of our team members, while taking a routine hourly EMF measurement, thought someone was talking to her from behind. On turning around, she found she was the only one in the area. And at the very end of the evening upon returning to our rooms, the same team member mentioned earlier again thought he heard boots on a hardwood floor in or around Room 307 but no one else was present.

The third night of the investigation mostly involved our experiments with those rolling balls, but a few things did occur which are worth mentioning. At one point early in the evening, one of our team members reported a similar experience to the one just described (but this time to a different investigator). While standing near and looking at a portrait of "Seth Bullock and the Cowboys," he turned around to talk to the team member who'd followed him into the room, only to find he was completely alone.

There were no other significant events for the duration of the evening, and we concluded our investigation, though we do have a couple of side notes of things we were able to figure out during "off" times when we weren't actively monitoring the location.

The "phantom clock" that sometimes rings even though the clock itself doesn't work was easy to solve by simply giving the clock a close examination. We determined that the reason the clock didn't work was because most of the interior mechanics had been removed. All that was left were the bell and the hammer that rings the bell. Because the rest of the mechanism was absent, the hammer was loose. Any small vibration, such as might be caused by a person walking in the room, could move the hammer and ring the bell. We're not going to say it's impossible for a ghost to do likewise, but the fact that simply walking through the room can trigger it means there's a far more parsimonious explanation.

The haunted mirror was also explainable. First of all, when photographed, it absolutely does contain an eerie sort of symbol on the mirror glass that doesn't seem to appear when viewed in person. People have described that symbol variously as a Celtic Cross, the anarchy symbol, "mysterious writing," and numerous other explanations. When we inquired about it of a staff member, we were told that it was one of the only remaining pieces of furniture from the original Hotel.

However, we also examined this mirror closely by very carefully removing the back board, where we found a store price tag still inside, proving that it was a more recent acquisition and not a true antique. This also gave us the explanation for the mysterious writing or symbol. It was the glue used to adhere the mirror to the cardboard backing. The glue had been applied in that shape, and while it's not visible to the naked eye, the camera's flash can penetrate the mirror's glass and cause the glue to shine through in a photograph.

So where does this all leave us? Clearly we were able to explain away a few of the alleged paranormal phenomena. Others, we weren't able to witness during our three days there, though it seems like at least some of them (such as the smell of cigar smoke or the sound of children crying) could be easily explained as merely the kinds of things that happen in hotels. When one is around strangers, one can expect some unusual sights, smells, and sounds. Even some of the things we did experience, like some of our team members thinking they heard someone when they were alone or the coffee machine deciding to turn itself on, while creepy in the moment, are just as likely to be the result of sound carrying from the active casino upstairs, so we aren't going to stake our reputations on calling those genuine evidence of paranormal phenomena.

But then there were those little balls. We never have been able to figure out either what caused the balls to roll or what caused the first video to disappear. Could it have been ghosts? We have to suppose it could be. Or it could be something we just haven't thought of yet. We're not prepared to call this proof positive of ghostly activity, but it certainly is one of the most intriguing cases we've worked on.

At the end of the day, despite a longer than average amount of time to investigate, we came away from this investigation with more questions than answers, and this one remains an open and unsolved mystery in our casebooks.

References & Further Reading

Bellanger, J. (2009). *Encyclopedia of Haunted Places: Ghostly Locales from Around the World.* Revised edition. Franklin Lakes, NJ: New Page Books.

Bullock Hotel (1993). The History of Sheriff Seth Bullock & The Hotel. Archived from the original on May 14, 2007 at <https://web.archive.org/web/20080509170642/http://www.heartofdeadwood.com/history/index.php>

Cornelius, J. (2024). The Finest Type of Frontiersman. *Frontier Partisans.* <https://frontierpartisans.com/34762/the-finest-type-of-frontiersman/>

Parker, W. (1981). *Deadwood: The Golden Years.* Lincoln, NE: University of Nebraska Press.

Wolff, D. A. (2009). *Seth Bullock: Black Hills Lawman.* Pierre, SD: South Dakota State Historical Society Press.

5

May This Room Remain Peaceful: Macky Auditorium

The Macky Auditorium Concert Hall, located at 1595 Pleasant Street in Boulder, CO, is part of the University of Colorado Boulder campus and is a multidisciplinary events venue operated by the University's College of Music. It's a great place to see a show and has hosted some of the world's most famous performers. It's also the site of a terrible murder which has prompted several reports of paranormal activity.

We have to admit, this is at once one of our favorite and least favorite cases. Favorite because the entire story is recent enough that it has allowed us to document the evolution of the paranormal lore in real time. And least favorite because it involves a real murder that touched the lives of people whose families still live in the area.

Figure 5.1. The Macky Auditorium. Photo: Bryan Bonner.

The History

Unlike in many of our cases, the history of the Macky Auditorium itself is actually fairly disconnected from the history relevant to the paranormal claims. We'll cover the entire history in brief, but it's a murder that took place at the Auditorium that will be relevant as we move forward to discuss the paranormal allegations and our investigation.

Our history begins with Mr. Andrew J. Macky, for whom the Auditorium is named. Though now featured in the Boulder County Business Hall of Fame, Mr. Macky did not come from the wealthiest of all possible backgrounds, and could be considered the epitome of a self-made man. He was born on November 1, 1834 in Herkimer County, New York and trained as a carpenter, eventually developing a reasonably successful lumber business in Wisconsin in 1857.[1] But by 1858, he set his sights on bigger prospects and relocated to Colorado as one of so many lured by the Pikes Peak Gold Rush, though he was never to find his success in gold. Instead, selling that business in 1859, he moved to Boulder where he worked, at various points in his career, as a carpenter (in which capacity he built Boulder's first frame building in 1860), postmaster, treasurer, justice of the peace, school secretary, court clerk, and first President of the First National Bank of Boulder, which he helped to found. Though autodidactic himself (Mr. Macky never attended college), he apparently had a great fondness for his local university. Upon his death in 1907, his will bequeathed no less than $300,000 (the equivalent of over $10 million in 2024 dollars) to the University of Colorado for the building of an auditorium, then the second-largest gift ever to a Colorado educational institution.[2]

Construction of the Macky Auditorium would take 13 years, beginning with a groundbreaking on September 20, 1909 and a cornerstone ceremony on October 8, 1910. The president of the University of Colorado engaged architects Gove and Walsh to design the building by incorporating elements of buildings—such as Italy's Palazzo Vecchio, England's King's Chapel and Magdalene Tower, and a Princeton campus building—with which he'd fallen in love during his travels. Though the architects did their job admirably, unfortunately the construction was delayed by legal battles as Macky's adopted daughter May, on finding she'd inherited nothing, sued the estate. When the matter was finally settled in the University's favor, construction was completed in 1922 with the addition of a $68,000 Austin pipe organ (with $20,000 approved by the Board of Regents and the balance raised through a local fundraising campaign). Macky Auditorium held its first concert on May 19, 1923.[3] Many people consider the CU Boulder campus to be one of the country's most attractive universities and the Macky is one of the

---

1   Boulder County Business Hall of Fame (n.d.) Andrew J. Mackey [sic] – Class of 1994. *Boulder County Business Hall of Fame.* <https://halloffamebiz.com/alumni/andrew-j-mackey-class-of-1994/> (accessed June 15, 2024).

2   CU Boulder College of Music (n.d.). Macky Auditorium Concert Hall: History. *University of Colorado Boulder.* <https://www.colorado.edu/macky/about/history> (accessed June 15, 2024).

3   *Ibid.*

most striking buildings on the campus.

In the decades since, the Auditorium has hosted a wide range of performers including University students and professionals in a variety of performing arts from around the world. It has also been the site of several cultural events, lectures, commencement ceremonies, pageants, conferences, college pranks, riots, and just about everything else one can think of.

Unfortunately, it's also the site of a tragic murder.

On July 9, 1966, twenty-year-old CU zoology major Elaura Jeanne Jaquette was eating her lunch on a patch of grass not far from the Macky, near an irrigation channel running between the Guggenheim and Hale buildings on campus, and enjoying some afternoon birdwatching while she waited for the children she'd been babysitting to get out of a nearby movie theater. The sequence of events that happened next is not perfectly clear, but the remains of Elaura's lunch, her binoculars, and her wallet were found where she'd been sitting.

Later that afternoon, police responded to a call that students had discovered a body in the isolated organ recital room in the Macky Auditorium's west tower. Unfortunately, the officers quickly determined the body was Elaura Jaquette, who had been brutally raped and murdered by a then-unknown subject. Indeed, the crime was so brutal that several of her teeth had been knocked out in the struggle. Perhaps in an attempt to mask her identity, her assailant had attempted to set her face on fire, but was unsuccessful as her blood prevented the fire from catching. The reason there was so much blood, police determined, was that Elaura attempted to crawl away from her attacker, who responded by swinging her by her feet until her blood splattered as high as twelve feet up most of the room's walls.[4]

Figure 5.2. The organ room as seen during our investigation. Photo: Bryan Bonner.

---

4   As part of our investigation, we obtained some of the police records, including some of the crime scene photographs, through public records requests. You don't want to see them.

About a month later, after interviewing more than 1,000 witnesses, police arrested a University janitor named Joseph Dyre Morse. According to the police records we were able to obtain, Morse, then thirty-seven years of age, was seen by one or both of his teenage daughters (or perhaps step-daughter as recorded in some of the police documents) returning home on the day of the murder carrying a bucket of bloody clothing which he then burned along with his own clothing in his incinerator.[5] A fingerprint found on a plywood board at the crime scene also matched one of Morse's fingerprints.

Though people generally thought of Morse as an unremarkable man, it later came to light that he apparently had an "inner beast" fueled by alcohol. Other (lesser) incidents of violence and drunken passes at women later came to light through witness testimony. Witnesses also reported seeing Morse drinking at a bar near campus the day of the murder. Though he pled not guilty at trial, Morse received a sentence of 888 years for first degree murder, one of the longest sentences passed in the state of Colorado.[6] He maintained his innocence throughout his incarceration until he made a terse (and, if we may editorialize, not very informative) confession in 1980. He died in prison in 2005 in his mid-seventies. As an interesting side note regarding the trial, the jury foreman, William S. Lovell, Ph.D., would later go on to study law and cited his experience serving on this jury as one of the motivating factors for his interest in the subject.

What remains unclear is how Morse got Elaura Jaquette from the grassy patch where she'd been having lunch up to the organ recital room. If he did so by force, he would have had to cover a fairly long distance into the building and then up a winding staircase. If he did so by persuasion, it's unclear why she left her belongings behind. Some have suggested that because Elaura was known to have been an accomplished singer and pianist, he may have somehow used her love of music to lure her to the organ room. Others think that because Elaura was a noted animal-lover, her attacker may have lured her by claiming there was some kind of critter in the organ room. This at least makes sense because it might explain why she was in such a hurry as to leave her belongings behind. Still, it's only speculation. According to Morse's otherwise unhelpful confession, he lured Jaquette into the organ room and then made sexual advances which she rejected and then, he said, "things got out of hand."

Some forty years later, the Jaquette family finally got some sense of closure when a memorial plaque was placed on a rock near where her belongings had been found back in 1966. The plaque bears a quote from poet Theodore Roethke: "It is neither spring nor summer; it is always." It was erected on what would have been Elaura's 61st birthday. The family said they want to honor and remember not her death but the life of the young girl who loved playing music in church and

---

5   A local historian told us this was the first case ever to make use of a closed-circuit television system in the courtroom so the young girls wouldn't have to be in the same room as Morse during the trial.

6   Dash, M. (2010). The longest prison sentences ever served. *A Blast From the Past*. <https://mikedashhistory.com/2010/07/24/a-prison-curiosity/> (accessed June 15, 2024).

wanted to become a biology teacher.[7]

Paranormal Claims

A variety of paranormal claims have sprung up in the decades since the murder at the Macky Auditorium. Unfortunately for the prospects of a thorough investigation, they mostly come in the form of vague claims often reported second- or third-hand, but they do fall into several distinct categories.

Many people over the years have told stories of a mysterious woman and/or a mysterious man in a brown suit wandering the Auditorium. Footsteps are said to be heard in unoccupied areas. Similarly, people say one can hear the sounds of talking, singing, or organ music when no one is present. Electrical appliances and lights are supposed to turn on and off of their own accord. Most disturbingly of all, supposedly blood stains from the murder reappear on the walls and floor of the organ room.

Our Investigation

While Macky Auditorium is often open to the public because it still functions as an events venue, the organ room where the murder took place, and which is supposedly the site of the various hauntings, is very much *not* open to the public. However, we were given access by the University to conduct a three-day investigation of that room over a weekend.

We arrived at about 9:00 p.m. on Friday evening and began our investigation following the standard protocol of establishing baseline readings on our various meters and determining locations for the monitoring equipment. Eventually we settled on establishing a base of operations on the east side of the main stairs opposite the door to the tower, though we had to move this on Sunday evening to the west wall of the stairs because a film crew occupied the second floor of the Auditorium on Sunday evening.

We established video monitoring at the following locations:
- On the landing just below our base camp (on Sunday this camera was moved to the door of the organ practice room),
- At the door of the tower looking back toward the base camp (on Sunday this camera was moved to the northeast corner of the stairs overlooking the new base camp),
- Ten feet from the door of the tower, pointed back toward the door,
- At the base of the stairs in the tower, looking up,
- At the top of the pipe room on the north side, looking down at the (original) location of the practice organ,
- At the top of the pipe room on the south side, looking down at the (original) location of the practice organ,

---

7   Daily Camera (2009). CU student murdered 40 years ago honored. *Daily Camera.* <https://www.dailycamera.com/2009/08/13/cu-student-murdered-40-years-ago-honored> (accessed June 15, 2024).

- On the bookshelf on the east side of the organ room,
- Outside the pipe room looking in,
- At the southwest corner of the organ practice room with a view of the entire room,
- At the door of the organ practice room, looking in, and
- At the top of the tower stairs, looking down.

Additionally, we placed microphones on the landing below our base of operations (which we moved on Sunday to the door of the organ practice room), at the base of the stairs, in the middle of the organ practice room, and on the bookshelf on the east side of the organ room. EMF meters were located on the bookshelf, on top of the practice organ, and in the middle of the practice room on a wooden chair. Thermometers were placed in the tower stairwell and at the base of operations. Balls used as control objects were placed in the same locations as the EMF meters. Both the balls and the EMF meters were within a camera's view at all times.

Finally, we used handheld EMF meters, AC/DC multimeters, seismometers, and handheld still and video cameras to measure and document the entire site throughout the three-day investigation.

Unfortunately, all that setup pretty much came to nothing as we didn't detect any unexplainable anomalies (video, audio, EMF, seismic, or otherwise) throughout the entire three-day investigation. But that doesn't mean we didn't still make some interesting discoveries.

One thing we found is that apparently it is (or at least was at the time of our investigation) a popular pastime for students to break into the location, probably on a dare to spend the night at the "haunted" room. We encountered several such groups of students, some of them even bearing sleeping bags and claiming to be "with the ghost hunters." We sent them all on their way and reported back to the University that this could be a problem they might want to discourage. Graffiti at the top of the stairs in the tower demonstrate that breaking into this area was a popular pastime.

We did detect some unusual EMF readings on Sunday evening, at relatively low levels but unusual compared to our baseline readings. However, when we heard thunder shortly thereafter, we determined that the momentary spikes in EMF were due to the atmospheric disturbances resulting from the nearby storm. Similarly, the building makes some loud creaks and pops due to expansion and contraction of its members, perhaps exasperated by the stormy weather. We also heard a lot of wind, passing cars, and sounds from radiators which we think could easily be mistaken (particularly if one is in a ghost-hunting sort of mood) for the sounds of music or human voices. Seismically the entire building exhibited a very low and consistent vibration which could potentially account for some paranormal experiences due to hallucination, but we note that this is speculative. We also found some variations in the building's electrical service that we think might be able to account for reports of electrical anomalies light lights turning on and off.

Because one of the paranormal claims involved people seeing blood stains from the murder even though the building had (of course) been cleaned and repainted since, we used a chemical compound called luminol to search for any remaining blood stains. Luminol is an agent that reacts with blood and causes it to luminesce a pale blue color. We used crime scene photos to determine where blood had originally been and applied the luminol to those surfaces. No blood was present. Because we conducted this experiment, we know that any blood stains reported now must have occurred after our investigation (whether from a natural or supernatural cause, one can only speculate).

For us, probably the most interesting part of our investigation into the Macky has been the ability to see how the story has evolved over the years. Several times since our initial investigation, when we've happened to be on or near the CU Boulder campus for whatever reason, we've asked around to find out what the students are saying about the paranormal rumors. What was fascinating to us was that the story changed over time. No one we ever spoke to claimed to have personally witnessed or experienced anything paranormal, but through tellings and retellings throughout the years, the story grew and became even more horrifying.

Though it was fascinating to see how the folklore evolved over time, it was also quite disturbing to us to see how a real-life murder developed into the campus equivalent of a campfire spook story. That sort of thing is par for the course in paranormal lore, but it was troublesome in this case because some of the victim's friends and family still live in the area. Fortunately, recent reports seem to suggest that the lore is beginning to die down and many students don't even know about the paranormal rumors anymore.[8] We'd like to think that perhaps the results of our investigation helped calm the rumor mill a bit.

Normally, we're all about the ghost stories, so it might seem out of character for us to actually want to completely dispel these rumors. The reason is quite simple. Not only did our investigation not find anything to suggest any kind of paranormal activity, but the friends and family of Elaura Jaquette have been quite clear that, due to Elaura's deep religious convictions, even if ghosts and hauntings are a real thing, she would not have been the kind of person to stay behind; she would have been the sort, they say, who would have forgiven her killer and been eager to be with her Lord. One family friend, speaking anonymously, said, "If ghosts are supposed to haunt the scenes of their demise because of the horrific nature of the crime that happened there, or their inability to accept their death, then that wouldn't have been her."

In the absence of any other evidence to the contrary, we're happy to accept that as the final conclusion on the supposed haunting of Macky Auditorium and to consider the case, at least unless someone comes up with something new, to be closed. We therefore conclude this chapter with the words of a graffito left on the wall at the entrance to the organ practice room: "R.I.P. Elaura Jacquette [sic]. May this room remain peaceful. Amen!"

---

8   Kerkhoff, G. (2023). A haunted history of CU's Macky Auditorium. CU Independent. <https://www.cuindependent.com/2023/10/31/a-haunted-history-of-cus-macky-auditorium/> (accessed June 15, 2024).

Figure 5.3. The Elaura Jaquette memorial graffito. Photo: Bryan Bonner.

References & Further Reading

Boulder County Business Hall of Fame (n.d.) Andrew J. Mackey [sic] – Class of 1994. *Boulder County Business Hall of Fame.* <https://halloffamebiz.com/alumni/andrew-j-mackey-class-of-1994/>

CU Boulder College of Music (n.d.). Macky Auditorium Concert Hall: History. *University of Colorado Boulder.* <https://www.colorado.edu/macky/about/history>

Daily Camera (2009). CU student murdered 40 years ago honored. *Daily Camera.* <https://www.dailycamera.com/2009/08/13/cu-student-murdered-40-years-ago-honored>

Dash, M. (2010). The longest prison sentences ever served. *A Blast From the Past.* <https://mikedashhistory.com/2010/07/24/a-prison-curiosity/>

Kerkhoff, G. (2023). A haunted history of CU's Macky Auditorium. *CU Independent.* <https://www.cuindependent.com/2023/10/31/a-haunted-history-of-cus-macky-auditorium/>

Lamb, K. (2016). *Ghosthunting Colorado.* Covington, KY: Clerisy Press.

6

Smoke and Fire: Onaledge Historic Lodge

Onaledge Historic Lodge, located at 336 El Paso Boulevard in Manitou Springs, Colorado, is so called because it's located on a ledge. It's a lovely estate near plenty of scenic attractions that now functions as a bed and breakfast. As it happens, it's also considered by some to be one of Colorado's most haunted hotels.

Our own investigation into the property was limited and brief but provides a good example of why we insist upon working with people from a wide variety of backgrounds because the solution to at least one of the paranormal claims came from a very specified body of knowledge.

Figure 6.1. Onaledge Historic Lodge. Photo: Bryan Bonner.

The History

We should begin by pointing out this is going to be one of the shorter chapters in this book. The history of the property is fairly limited, as far as we've been able to discover. And, in fact, our entire case file on this one is fairly thin. Here's what we do know.

The Onaledge house was built in 1912 by an English coppersmith named Roland Boutwell. It was later purchased by one Mr. Miles Frank Yount, owner of the nearby Rockledge estate, for use as a guesthouse for his growing family who liked to summer at Rockledge. It's currently run as a bed and breakfast.

Despite our attempts to locate interesting history at the property, we've largely come up empty. Some have claimed that Onaledge is listed in the National Register of Historic Places, which would certainly suggest some interesting historical significance.[1] Unfortunately, while it is certainly a historic property in the sense of its age and Arts & Craftsmanship architectural style, and while it is located in the historic district of Manitou Springs, our searches of the State and National registers of Historic Places likewise came up empty.

That's not to criticize the property, by any means. Few would argue that it's a remarkable building. Nor is it even to say that nothing interesting happened there. We maintain that even the most ordinary of people are worthy of being remembered by history. Something fascinating happens to everyone. What we are saying, though, is that it unfortunately seems to be the case that much of the history of Onaledge, much like so much other history, has been insufficiently documented or preserved for future study.[2]

---

1   DuVal, L, & Banks, M. (2011). Insiders' Guide to Colorado Springs. Guilford, CT: Globe Pequot Press.

2   On an interesting side note that has nothing whatever to do with Onaledge but is relevant to this topic of preserving even the most ordinary of all histories, we'd like to call your attention to a wonderful initiative called "The Great Diary Project," housed at England's Bishopsgate Institute. Founded by Drs. Polly North and Irving Finkel (the latter of whom is a noted expert on ancient Mesopotamian cultures and wrote a wonderful book on the earliest recorded ghost lore called *The First Ghosts*, which we heartily recommend for people interested in scholarly work within the paranormal), the project is collecting and preserving as many diaries as they can get their hands on so future scholars can read the daily writings of the kinds of ordinary people we've been talking about. The project began in 2007, found its permanent home at Bishopsgate in 2009, and as of this writing in 2024 has rescued and made available to researchers more than 19,000 unpublished diaries, making it the largest collection of its kind of which we're aware. An American counterpart (albeit so far with a much smaller collection), was founded in 2022 by Kate Zirkle and is called "The American Diary Project." We lend our absolute support to initiatives such as these which help to preserve the kind of history we so often struggle to locate in our research.

Paranormal Claims

The haunted reputation of Onaledge is undoubtedly better-known than its history, having received even national press attention as one of Colorado's most haunted hotels.[3] However, though its reputation is well-documented, the particular claims, lacking the kind of historic basis found in some of our other case files, are often fairly vague.

Some of the spirits we've seen reported as haunting Onaledge include a "Grandpa" character who has been seen reading to a little boy in a blue suit by the fireplace, a Victorian couple, a Victorian woman (who may or may not be the same individual as reported to be half of the aforementioned couple), and a young girl perhaps about eight years of age, as well as unspecified others, with the total number of spirits supposedly somewhere between about five and about a dozen.[4]

The thing that may be a bit unusual about the alleged haunting of Onaledge is that an uncharacteristically high proportion of the paranormal claims involve visual manifestations of the spirits. Often people report hearing things, disturbances with appliances, electromagnetic anomalies, and all manner of other things, but at Onaledge, the majority of claims seem to involve people actually seeing full-bodied manifestations of the spirits themselves.

However, the one that caught our attention and sparked our own investigation was a video, apparently from a surveillance or security camera, taken at Onaledge in which a mysterious light quickly flashes across the screen. Though we can't reproduce the image, both because print can't contain video and out of respect for copyright, we can describe it. The video is a black and white shot from a camera that appears to be mounted either on the ceiling or on top of some tall structure. The top of the frame shows a ceiling fan, and the camera is pointed across a room with a bed at the far side. In the middle of the footage, a luminescent subject, quickly flashes across the screen and then disappears not off the side of the frame but right in the middle of the camera's field of view. Unlike the witness accounts we've described above that report identifiably humanoid spirits, this one appeared to just be a sort of flash of light.

Our Investigation

Our investigation into the hauntings at Onaledge was pretty quick and limited. None of us have ever experienced or witnessed the vast majority of the paranormal phenomena reported there, so there's really not a whole lot we can say about such alleged spirits as the Grandpa and the young boy or the Victorian woman. We're not saying they do or don't exist; they're just not something we've personally experienced.

---

3   Brady, P. (2012). Onaledge, Manitou Springs, Colorado's Very Haunted Hotel. Huffpost. <https://www.huffpost.com/entry/onaledge-manitou-springs-haunted-hotel_n_1982265> (accessed June 16, 2024).

4   *Ibid.*

But that video of the flash of light caught our attention, and we were able to determine a plausible source of that one. And this is actually what makes paranormal investigation so interesting to us, because one never knows what field of expertise will suddenly become relevant. In past investigations, we've called upon psychologists, geologists, chemists, anthropologists, and more. But in this case, it was some of our team members' background and training in magic (which is to say conjuring tricks, not "real" magic or occultism) that solved the case.

Upon reviewing the video a few times, it appeared like a flash of fire moved through the room. But there was no ash or smoke. We were able to recognize that the flash of light, the speed of combustion, and the lack of smoke and ash, looked a lot like what our magicians have seen when working with a substance known as "flash paper."

More properly known as nitrocellulose, flash paper (and a variety of other "flash" products like flash string and flash cotton) is made by treating tissue with a mixture of concentrated nitric and sulfuric acids. It was actually discovered by accident when a chemist used a cotton towel to mop up a spilled mixture of those acids and the heat from the outside of his stove caused the towel to rapidly combust. That ought to demonstrate that this is a volatile substance. In fact, the reason it produces very little smoke or residue is partly because of the rapidity of its combustion.

Magicians and other trained performers can use commercially available flash products safely (and it's even fairly easy to make your own but we will *not* tell you how because we don't want to encourage foolish behavior and making flash products safely should really be done in a professional laboratory equipped with a fume hood and proper safety devices[5]) despite the volatility of the product.

The important thing in terms of the alleged ghost video is that the behavior of the supposed spirit looks identical to that of flash paper tossed in front of a camera. It burns brightly and rapidly (it's called *flash* paper for good reason) and disappears in an instant leaving nothing noticeable behind.

Though we didn't recreate the video on site at Onaledge (we're not in the habit of intentionally starting fires—even relatively safe and controlled ones—at someone else's property), we did film several recreations of the video in our own homes, a still image from one of which is reproduced below.

Based on that recreation, we're reasonably convinced the Onaledge video was actually the result of someone standing out of frame tossing lit flash paper into the camera's view. That leaves us in a bit of an awkward situation because the vast majority of paranormal claims we've investigated, even if we're able to supply natural explanations, are not the result of hoaxes. Contrary to what peo-

---

5   Seriously, don't mess with the stuff if you don't know what you're doing. Making it involves working with concentrated strong acids. We've heard of multiple people who've died while attempting to manufacture flash products. Even storing it can be somewhat dangerous, and we've heard stories from other magicians about flash paper simply stored in a drawer or filing cabinet spontaneously combusting and destroying homes because the autoignition temperature is so low. For this reason, even professionally-made flash products purchased from reputable retailers should always be stored wet in a cool dark place (and not for too long a time) as a safety precaution.

ple seem to think, hoaxes are actually pretty rare in our experience. But this one certainly has the elements of a hoax.

Figure 6.2. Recreation of flash paper image. Photo: Robert Lewis.

Importantly, though, we're not accusing anyone of anything in particular. While we believe the video to show flash paper rather than a ghost, which would mean someone intentionally created the phenomenon, we don't know that any-one intended to intentionally pass it off as a ghost. The video's provenance is unclear to us and we don't know who shot it or for what reason it was shot, so it's entirely possible that the video was created for innocent purposes and misunder-stood and misattributed by other people after the fact. Of particular importance, we're not saying the management of the Onaledge Historic Lodge have or had any knowledge of or participation in the creation of the video, whether it was an intentional hoax or not. We simply don't know why the video was made.

Despite the brevity of this case (and this chapter), we do like this one. For one thing, Onaledge really is a lovely property. Ghosts or no, it'd be a great place to stay for a weekend. But more on topic for us is observation, demonstrated here quite clearly, that investigating anomalous phenomena requires a willingness to look to unexpected sources of expertise. Had some of our members not trained as magicians, there's no telling how long it might have taken us to recognize this video as likely being flash paper. The important lesson there is to always approach these investigations with the necessary humility to say "I don't know what that is, and I'm going to ask everyone I can possibly think of until someone can identify it." Our own backgrounds made this one easier for us than it might have been for other teams, but the converse is also true—there are cases in which a perfectly simply explanation might elude us simply because we lacked the training in some obscure field to recognize a phenomenon for what it was.

References & Further Reading

Brady, P. (2012). Onaledge, Manitou Springs, Colorado's Very Haunted Hotel. *Huffpost*. <https://www.huffpost.com/entry/onaledge-manitou-springs-haunted-hotel_n_1982265>

DuVal, L, & Banks, M. (2011). *Insiders' Guide to Colorado Springs*. Guilford, CT: Globe Pequot Press.

7

A Haunted Library: The Sarah Platt Decker Branch Library

The Decker Library (or more properly, the Sarah Platt Decker Branch Library), located at 1501 South Logan Street in Denver is a small but charming branch of the Denver Public Library system whose homey feel begins with the building's house-like exterior and extends to the welcoming interior. It's located at the border of Platt Park, making it a wonderful place to check out a book and spend the day reading in a park.

It also has a haunted reputation among both patrons and staff. Which makes sense to us. If we were to become ghosts and haunt any location, a library would probably top our list of selections. Even as living humans, the opportunity to spend some time in a haunted library was too good to pass up, making this brief investigation one of our most enjoyable.

Figure 7.1. The Sarah Platt Decker Branch Library. Photo: Bryan Bonner.

The History

The history of the Sarah Platt Decker Branch Library is intimately connected with the histories of Denver itself, the Denver Public Library system at large, and the woman for which it was named, Sarah Platt-Decker.

When the first settlers established homes in Denver, no one expected the town to become the large city it eventually developed into. But as the small pass-through settlement started evolving into a true city, the need for a library system became clear. Indeed, the movers and shakers responsible for the creation of the library system include the very same names that come up time and again in any local history book. Though an earlier library had been established in Denver as early as 1860 operating under a paid subscription model, the first attempt at a public library system was the Denver Library Association founded in 1874 by several civic-minded businessmen including none other than Walter Scott Cheesman (about whom you can read in much greater detail in our chapter on Cheesman Park in Volume 1, Chapter 5).[1] Unfortunately, due to lack of financial support, this initiative failed after just four years and the Association's collection of books were donated to the local school system.

The current Denver Public Library system was founded shortly thereafter in 1884 by the local Chamber of Commerce who initially provided all financial support, though public support was approved by the Denver City Council in 1891, eventually becoming fully supported and operated by the City in 1898. From 1910 through 1955, much of the private support for public libraries in the United States—and certainly including the Denver Public Library system—came from Scottish-American industrialist and philanthropist Andrew Carnegie (1835-1919) and, after his death, his estate and his various foundations. Mr. Carnegie contributed to the opening of more than 2,500 public libraries including 36 in the state of Colorado.[2] The Decker Library was the second (after the Ford-Warren Branch) of four Carnegie-funded libraries in Denver.

Funding for the Library came from a Carnegie grant approved on March 14, 1902. Construction began in 1911 and the Branch opened for business on June 17, 1913.

Its namesake, Sarah Sophia Chase Platt-Decker, born Sarah Sophia Chase in 1856 in Vermont, was a noted local suffragist and philanthropist. She arrived in Denver in 1887 with her second husband (her first was Charles B. Harris, who died two years into the marriage), James H. Platt, Jr., a physician and former United States Congressman. Both members of the couple were active in local politics and civic concerns, and the park adjacent to the library was named in Mr. Platt's honor. After Mr. Platt died in 1894, Sarah was married a third time to Mr. Westbrook S. Decker, a Denver judge. When Decker died in 1902, Sarah Platt-Decker did not remarry. Upon her own death in 1912, Denver City and County offices closed early (at noon) and ordered flags to be flown at half-mast. The pallbearers

---

1   Denver Public Library Special Collections and Archives (n.d.). History of the Denver Public Library. Denver Public Library. <https://history.denverlibrary.org/exhibit/history-denver-public-library> (accessed June 18, 2024).

2   *Ibid.*

at her funeral included three Colorado governors.[3]

Why was her legacy so honored? She contributed to the local community through a number of political and philanthropic initiatives including charitable work with children and hospitals, but is best known for her work as a suffragist and women's advocate. That cause began upon the death of her first husband when she received only a "widow's third" of his estate, which was also the reason she did not retain his surname. Among her accomplishments in this line of work were the founding in 1894 of the Woman's Club of Denver (of which she served as the first president), service in the presidency of the General Federation of Women's Clubs from 1904 through 1908 (representing two terms, during which she's credited with expanding the organization's national prominence), and probably most significantly, influence in the 1893 vote that won women's suffrage in the state of Colorado (making Colorado the second state in the Union to pass such legislation). She was also instrumental in the formation of the Mesa Verde National Park in 1906, after which her work in conservationism earned her an invitation from President Theodore Roosevelt (about whom see also Chapter 4) to attend the Governors' Conference on Conservation of Natural Resources at the White House in 1908.[4]

The library itself was designed by architects Marean and Norton who also designed a large number of other Denver buildings, both private residences and public landmarks, including the Cheesman Park Pavilion (again see Volume 1, Chapter 5, in which we contrast the remarkable architecture planned by Marean and Norton with the unfinished work the City actually produced). The Decker Library stands out among libraries for its L-shape and English domestic style, featuring a green tile roof and brick walls and chimneys extending from a large fireplace beneath a Dudley Carpenter painting of the Pied Piper of Hamelin, giving it the appearance arguably more of a home than of a public institution.[5]

The building was designated a historic landmark by the City and County of Denver in 1984[6] and assigned landmark number 155.[7] After a period of falling

---

3   *Ibid.*

4   Duncan, E. (n.d.). Sarah Platt Decker. Colorado Encyclopedia. <https://coloradoencyclopedia.org/article/sarah-platt-decker> (accessed June 18, 2024).

5   Denver Public Library Special Collections and Archives (n.d.). History of the Denver Public Library. Denver Public Library. <https://history.denverlibrary.org/exhibit/history-denver-public-library> (accessed June 18, 2024).

6   City and County of Denver (2023). Denver Individual Landmarks. City and County of Denver. <https://www.denvergov.org/files/assets/public/v/4/community-planning-and-development/documents/landmark-preservation/individual_landmarks_list.pdf> (accessed June 18, 2024).

7   Or perhaps number 153. The official list of "Denver Individual Landmarks" published by the City and County of Denver lists the Library as number 153 in its "running list" as of the most recent publication in 2023 (see citation above). However, the plaque erected on the building upon its designation lists it as landmark number 155. This discrepancy is most likely due to two designated sites appearing earlier on the list (Constitution Hall and the Thomas M. Field House) being designated sites even though the buildings themselves were demolished. These sites are listed in the "Denver

into disrepair, the building was restored to its former glory in 1993 and remains open for business to this day.

Figure 7.2. The reading room with the fireplace and painting. Photo: Bryan Bonner.

Paranormal Claims

A variety of paranormal claims have been reported at the Decker Library over the years by both patrons and staff. Though the building's history is well-documented, none of the alleged ghostly phenomena have (as far as we can tell) been attached to any particular historic figures. One might expect the ghost of Sarah Platt-Decker to haunt the building that bears her name. Conversely, one might expect her spirit very much not to haunt the building because it was only opened after her death and so she never actually spent any time there. Lore we've encountered over the years would suggest either possible interpretation: that spirits haunt places only where they personally lived and/or died, or that spirits can

---

Individual Landmarks" document but are not numbered in the running list. A similar omission from the running list concerns the Central Bank (West) building which was designated in 1988 but demolished and de-designated in 2008 (though it occurs below the Decker Library in the document and so doesn't affect the Library's numbering). It's unclear, therefore, precisely which numbering system the City and County of Denver uses to track these historic designations. We've chosen to use the designation number 155 in the main text both because it's the number on the building's plaque and a prior version of the "Denver Individual Landmarks" document, though we would certainly prefer if the City would correct the discrepancy. When we reached out to call their attention to the discrepancy, however, they indicated that they no longer number landmark designations (despite numbering them in their running list), so it appears they're comfortable with the inconsistency as it stands.

move about and haunt any building with which they might feel some kind of personal connection. As usual, we remain neutral on these points of paranormal philosophy because there's no solid evidence to support either interpretation.

Regardless of who it might be, though, many people believe there is/are some spirit(s) haunting the Decker Library. For us, that seems almost too good to be true. Haunted libraries seem like such wonderful places, and at least some of our members have expressed a life goal of someday living in a historic mansion with an elegant library haunted by a whimsical library ghost. Seems unlikely to happen, but one can always dream.

Anyway, the majority of paranormal claims surrounding the Decker Library involve claims of spirits manipulating objects within the library.

Numerous reports suggest that the Library's front door opens seemingly of its own accord. This occurs even if the door is locked after business hours. Conversely, on other occasions, the door is reported to have locked itself even *during* business hours and with nobody near the door. Doors opening and closing are, of course, common claims of supernatural or spirit activity. Often, though, these can be easily explained as the result of something like a draft blowing the door. In this case, because the door is said not only to move but to lock and unlock itself, it seems like a more intriguing claim.

Speaking of things happening after business hours, several people have reported seeing people moving about inside the building even when it's closed and locked.

Any good library haunting needs some paranormal action with the books. The Decker Library is no exception. Multiple witnesses have said the ghosts have knocked books off the shelves. Again heading off a potential mundane explanation, they've pointed out that the shelves are angled slightly toward the back of the bookcases. That is, if something were to fall simply because it was precariously placed, it ought to fall onto the shelf, not off of it as these books have been reported to do. Some of the witnesses have reported an even more impressive feat of the books "jumping" from the shelves.

Some balls for the children are kept on a set of shelves behind the main counter. Though these shelves have the same "reverse angle" design, the balls have reportedly rolled off the shelves in a feat reminiscent of our intriguing and frustrating experiences described in Chapter 4.

A meeting room in the Library's basement has several haunting stories. Sounds of people in the basement even when no one is supposed to be there have been reported. But our favorite story from the basement is that a piano stored in the meeting room sometimes plays by itself even when the basement is unoccupied.

Finally, any haunted location worth its salt is going to have ghosts messing with the electrical devices. In this case, a security guard reported that on two separate occasions the light fixture in the restroom would fall from the ceiling when he was alone in the room. Usually the ghosts are said to just turn the lights on and off, so this was a bit more exciting than the average tale.

Figure 7.3. The basement with the piano. Photo: Bryan Bonner.

Our Investigation

The Library's management reached out to us several years ago to inquire about arranging a combination of a paranormal investigation with a lecture to follow. That's always a fun arrangement for us because not only do we get to spend some time doing our thing in the haunted location but we also get to publicly share our findings in the very same location we just investigated. Plans were made and we conducted a one-night investigation at the Decker Library.

Unfortunately, this particular investigation didn't yield too many interesting results. After a thorough tour of the property, we started off the evening by setting up our usual combination of video, audio, EMF, and seismic monitoring. During the tour, we did note that the "reverse angle" of the bookcases had been correctly described to us, so the claim that objects should fall into rather than off of the bookshelves unless propelled by some outside force makes sense.

Throughout the evening, we detected some abnormally high EMF readings. However, we were quickly able to determine their source: Wi-Fi equipment installed throughout the library created the stronger than normal electromagnetic fields.

It's true that for some people and under some circumstances, strong electromagnetic fields (while not normally a health risk) can cause hallucinations of paranormal phenomena. Not only is this well-documented in the scientific literature but we've witnessed it personally on some investigations (the most noteworthy of which is slated for publication in a future volume of this series). In this case, though, we don't necessarily think this is a good explanation for the types of claims that have been reported. Perhaps one or two reports could have been caused by these EMF-induced hallucinations (though we have no direct evidence to think this is the case), but it would be quite extraordinary for all of the various claims by various witnesses to share this as a common solution.

But that's unfortunately where we have to leave things because other than the quickly-explained elevated EMF readings, we didn't witness or experience anything anomalous during our night at the Decker Branch Library.

Of course as we always remind people, our failure to document these phenomena during our visit doesn't mean they didn't occur. It simply means they didn't occur in our presence, so we have to leave this case open and unsolved. We can speculate about some of the claims—for example, the reports of people being seen in the building after hours could hypothetically have merely been clean-ing staff or an after-hours event of some kind—but unless and until we witness the phenomena for ourselves, we can only report what others have told us.

References & Further Reading

City and County of Denver (2023). Denver Individual Landmarks. *City and County of Denver.* <https://www.denvergov.org/files/assets/public/v/4/community-planning-and-development/documents/landmark-preservation/individual_landmarks_list.pdf>

Denver Public Library Special Collections and Archives (n.d.). History of the Denver Public Library. *Denver Public Library.* <https://history.denverlibrary.org/exhibit/history-denver-public-library>

Duncan, E. (n.d.). Sarah Platt Decker. *Colorado Encyclopedia.* <https://coloradoencyclopedia.org/article/sarah-platt-decker>

8

A Haunted Train Station: Denver Union Station

Denver's Union Station, located at 1701 Wynkoop Street, is the city's main railway and transportation hub. It's both a major station in the heart of Denver to serve the day-to-day public transportation needs of the city, including the RTD Light Rail system, as well as the place people go if they're planning a journey across the nation by passenger train. Of course, like any old public structure, it has its share of both history and ghost stories. Of the latter, some seem directly connected to events from history while others seem simply to have attached themselves to the Station for reasons unknown. Unfortunately, because Union Station remains an active transportation hub, our own investigation has been necessarily limited in scope.

Figure 8.1. Denver Union Station. Photo: Bryan Bonner.

The History

Readers who've been paying attention to our books will know that Denver was never originally intended to be the major city it eventually became. Early settlers in the region were often just passing through on their westward journey as part of a gold rush. Locals even sometimes joke that the very existence of Denver (and indeed the entire Denver Metropolitan Area encompassing Denver proper as well as all the nearby suburbs) is a testament to laziness; that pioneers moving west reached the foot of the Rocky Mountains, immediately said "hell no," to crossing any further, and just settled where they were. Of course that's not the reality of the situation, but any good joke does contain a hint of truth. The reality is that lots of people settled in the Denver area because there was plenty of mining right here in Colorado. But because for many people it was initially a stop on a longer journey rather than a final destination, it began largely as a pass-through settlement, then a pass-through town, and only gradually developed into a city.

At about the same time Denver was beginning to develop into a full-fledged city, railroads were being constructed both to service long-term Denver residents and to carry passengers and goods across the nation. Though the "gateway to the west" moniker is more properly applied to St. Louis, it's nevertheless a fitting description of Denver and its place in the growing rail system.

Initially, there was no central hub for all the rail companies. Beginning in 1868, several different rail companies including Union Pacific, Kansas Pacific, and others each operated independent train depots in Denver. However, the need for a single depot quickly became apparent as the separately-operated facilities caused difficulty and confusion when passengers who needed to move from one line to another found themselves being shuffled off to entirely different stations. In 1879 several prominent men held a meeting to form the Union Depot and Railroad Company with none other than Walter Scott Cheesman (see Volume 1, Chapter 5) as its first President. Shortly thereafter, in February of 1880, they secured contracts with the four rail lines who would use the new "Union" depot.[1]

During the course of our research into the history, a local historian told us a story (perhaps better classified as a rumor) which we haven't found the documents to verify but which is just too good not to share. Apparently one of the small train depots (before the unification) which was located just east of the current Union Station location had a problem with people failing to use the spittoon in the building's lobby. Their first attempt to correct the problem was to place a sign in prominent view which said, "Please do not spit on the floor—use the spittoon" (or words to that effect). When that failed, the frustrated powers that be got a novel idea. They somehow "acquired" a human head from the City Cemetery (now Cheesman Park—see Volume 1, Chapter 5 for a detailed account of what became of the rest of that cemetery's residents) and placed it next to their

1   Stevens, M. E. (1974). Union Station: National Register of Historic Places Inventory – Nomination Form. [Register No. 74000571]. *National Register of Historic Places.* <https://npgallery.nps.gov/GetAsset/47425798-482f-4ad0-8984-f3ae30a1b7dc> (accessed July 2, 2024).

original sign along with a new sign that added "this is the last guy who spit on the floor." Supposedly this solved the problem. When the building was demolished years later, according to the rumor, the head was buried where Union Station is today. Can we verify this rumor? Not even close. And it would certainly be an extraordinary tale if true. But you know us by now. If there's a creepy story, we almost have no choice but to repeat it.

Returning to the verifiable history, the unified station, then called the Denver Union Depot, opened on June 1, 1881, boasting a 500-foot-wide building and an ornate 180-foot-high clocktower, making it at that time the tallest building in the West.[2] It was designed by architect A. Taylor of Kansas City in the Italian Romanesque style and featured a volcanic stone foundation and pink-grey rhyolite walls, both quarried from Castle Rock.[3]

It was not to last. Recall that we pointed out in Chapter 1 (and elsewhere) that pretty much the entire city of Denver (indeed, pretty much the entire state of Colorado) burned down at some point. The original Union Depot was no exception. On March 18, 1894, less than three years from its opening, a fire caused by an electrical short obliterated the central portion of the station. Though insurance paid the claimed loss of $125,000 (equivalent to just over $4.5 million in 2024), the actual reconstruction would end up costing an additional $75,000 (or just over $2.8 million in 2024 dollars). Reconstruction began promptly following a design by architects VanBrunt and Howe (also of Kansas City). Though taller than its predecessor and featuring four clocks on its clocktower, historians consider the new design "less distinguished" than the original.[4]

Just outside of Union Station, and dedicated on Independence Day of 1906, there was an archway known either as the Denver Welcome Arch or the Mizpah Arch. It initially had the word "welcome" displayed prominently on both sides, but when it was thought it odd to bid people "welcome" as they were *leaving* the city, a change seemed to be in order. Yet they didn't want to say "goodbye," because that sounded too ominous. Instead, opting for a poetic sort of farewell statement, they changed one side to the Hebrew word *Mizpah*, which literally translates as "watchtower" but was a reference to a quotation in Genesis which reads "And [the pillar was called] Mizpah; for he said, The Lord watch between me and thee, when we are absent one from another."[5] It received electric lights on its lettering in the 1920, making it arguably even more delightful, but was ultimately torn town by the City in 1931 as a traffic hazard.[6]

2   Union Station (n.d.). Our History: A Century and a Half in the Making. Denver Union Station. <https://www.denverunionstation.com/about/our-history/> (accessed July 2, 2024).

3   Stevens, M. E. (1974). Union Station: National Register of Historic Places Inventory – Nomination Form. [Register No. 74000571]. *National Register of Historic Places*. <https://npgallery.nps.gov/GetAsset/47425798-482f-4ad0-8984-f3ae30a1b7dc> (accessed July 2, 2024).

4   *Ibid.*

5   Genesis 31:49 [KJV].

6   Walden, S. (2020). Denver's Welcome Mizpah Arch. *Colorado Railroads*. <https://www.corailroads.com/2020/03/denvers-welcome-mizpah-arch.html> (accessed July 2,

But even the newly rebuilt structure didn't last very long. The Union Depot and Railroad Company began to disband in 1912, being replaced by the Denver Terminal Railway Company.[7] The name was changed to the Union Station and the central structure of the depot was again razed and replaced by the Beaux-Arts style building that still exists today.[8]

Early in the morning of August 3, 1933, disaster again struck. Following days of heavy rainfall, the Castlewood Dam, located about thirty miles southeast of Denver and which had stood for 43 years, suddenly failed, unleashing a devastating torrent of water that would kill two people and damage thousands of properties including flooding Union Station with six inches of water for several days.[9] Union Station again found itself (along with much of the state of Colorado) under water in the infamous South Platte Flood of 1965 which claimed twenty-four lives and left some areas submerged under as much as three feet of water.[10]

In the mid-20th century, and particularly following World War II, Union Station began to fall into neglect and disrepair. Air travel largely replaced the rail system as the popular choice for most people in most circumstances (though now in the early decades of the 21st century, some people are beginning to look again to rail travel as an alternative to the inconveniences of air travel, with varying degrees of success). But in 2001, Denver's Regional Transportation District (RTD) purchased Union Station and began a period of reconstruction. The revitalized Union Station, now featuring not only access to local and interstate railways but a shopping and cultural district along with bars, restaurants, and even a hotel, reopened in 2014, marking 100 years in its current building.[11] It now features twenty-two bus gates, eight commuter rail tracks, three RTD Light Rail tracks, and a total of seven rail platforms.

2024).

7   Stevens, M. E. (1974). Union Station: National Register of Historic Places Inventory – Nomination Form. [Register No. 74000571]. *National Register of Historic Places*. <https://npgallery.nps.gov/GetAsset/47425798-482f-4ad0-8984-f3ae30a1b7dc> (accessed July 2, 2024).

8   Union Station (n.d.). Our History: A Century and a Half in the Making. Denver Union Station. <https://www.denverunionstation.com/about/our-history/> (accessed July 2, 2024).

9   Garrison, R. (2023). Castlewood Dam failure, 90 years later: The day Denver was hit with a wall of water. *Denver 7*. <https://www.denver7.com/news/digital-originals/castlewood-dam-failure-90-years-later-the-day-denver-was-hit-with-a-wall-of-water> (accessed July 2, 2024).

10   Prendergast, A. (n.d.). South Platte Flood of 1965. *Colorado Encyclopedia*. <https://coloradoencyclopedia.org/article/south-platte-flood-1965> (accessed July 2, 2024).

11   Union Station (n.d.). Our History: A Century and a Half in the Making. Denver Union Station. <https://www.denverunionstation.com/about/our-history/> (accessed July 2, 2024).

Paranormal Claims

With such a long history, including several disasters and a rumored severed head, it's probably surprising to no one that Union Station has its share of ghosts. Or at least of ghost stories.

Probably the most prominent is the story of a headless body wandering the property. Both guests and staff have reported seeing this specter in various locations throughout the building. This builds upon the rumor that a severed head (used to encourage people to make use of the spittoon in the prior train depot) might be buried somewhere on the property. The story goes that the night watchman of that former depot even saw this spirt before Union Station existed. The ghost wandered throughout the building making odd sounds and moaning. Presumably the ghost was looking for his head. If so, and depending on how such things work in the spirit world, he might be in luck. Other witnesses have also seen a ghostly head without its body floating around Union Station.

Another night watchman, working at a more recent incarnation of Union Station, said he was working alone late one evening after everyone else had left the building.[12] While he was doing his rounds, he heard what he could only describe as a "large party" coming from a locked storage room in the basement. He went to investigate and the sound got progressively louder as he approached the locked door. But as soon as he opened the door, the sound immediately stopped. He was never able to find the sound's source.

Two other spirits have been reported by several witnesses. One of them is a "three-fingered[13] hobo" who used to live at the station. It's said he likes to bother the ticket agents and passengers. The other is a little girl dressed in 1800s attire who has been seen haunting the tunnels. Some have speculated she might have been a victim of the 1894 fire.

Others have reported ghostly activity that didn't visibly manifest as a full apparition. Several people reported feeling a "strong" or "militaristic" presence which they attribute to a ghost whose identity is not known. Perhaps a military man who passed through Union Station to board a troop train during one of the wars? Or maybe a guard or someone else in a position of authority at the train depot? Other employees have described a variety of phenomena including a telephone dialing 911 of its own accord, papers disappearing from a desk, and 1930s or 1940s music playing in an abandoned men's restroom.[14]

---

12   Obviously this part can't be entirely correct. Supporting not only a major transportation hub but also bars, restaurants, and a hotel, there's no way for the Union Station complex to ever be entirely empty of people. However, we assume he meant that the *part* of the complex he was patrolling was vacant, which is entirely believable as plenty of the individual areas do indeed close after normal business hours.

13   It's not clear to us whether this means he has three fingers on one hand, three fingers on each hand, or three fingers total.

14   RTD FasTracks (2014). Haunted Denver Union Station. [YouTube video; channel: RTD FasTracks]. *YouTube*. <https://www.youtube.com/watch?v=1uWEiTrG1E8> (accessed July 2, 2024).

Our Investigation

Unfortunately, there's really not much to say here so this is probably going to be the briefest of all of our investigation reports. Because Union Station is such a bustling facility and responsible for so much rail and bus travel, we had no way to shut the entire place down to conduct a proper investigation according to our usual methods.

The "Great Hall," what (aside from the tracks themselves) most people think of as "the" Union Station now exists in a sort of public-private cooperation. It's publicly funded and has even been called "Denver's Living Room" but is leased by the privately-held Denver Union Station Alliance and serves as the main lobby for both the Crawford[15] Hotel (private) and the RTD station (public). Though its renovation was paid for by the taxpayers, they now prohibit video or audio recording within the Great Hall and even station guards at the restrooms and limit access to paying customers only (presumably in an attempt to crack down on high crime rates).[16]

With all that going on, there really hasn't been much we could do except spend a lot of time quietly watching (but of course not recording) from the publicly accessible locations of Union Station. After years of doing that, we haven't seen anything paranormal for ourselves. We even second guessed whether we ought to include Union Station in our case files for that reason, but finally decided to do so on the grounds that the ghost stories and history are interesting, even if our own activities aren't.

At the end of the day, Union Station is a wonderful place to explore, with a fascinating mix of historic structures and modern amenities. But is it haunted? That's anybody's guess.

References & Further Reading

Garrison, R. (2023). Castlewood Dam failure, 90 years later: The day Denver was hit with a wall of water. *Denver 7*. <https://www.denver7.com/news/digital-originals/castlewood-dam-failure-90-years-later-the-day-denver-was-hit-with-a-wall-of-water>

Harris, K. & Beaty, K. (2022). Is Union Station really dangerous? DPD's crackdown mostly busted drug users. *Denverite*. <https://denverite.com/2022/05/13/denver-union-station-arrests-tickets-numbers-violent-crime-drugs-addiction-treatment-homelessness-mayor-michael-hancock/>

Prendergast, A. (n.d.). South Platte Flood of 1965. *Colorado Encyclopedia*. <https://coloradoencyclopedia.org/article/south-platte-flood-1965>

---

15   Named in honor of, but not founded by, Dana Crawford, who was responsible for much of Denver's historic preservation and revitalization. Read more about her work in Volume 1, Chapter 2.

16   Harris, K. & Beaty, K. (2022). Is Union Station really dangerous? DPD's crackdown mostly busted drug users. *Denverite*. <https://denverite.com/2022/05/13/denver-union-station-arrests-tickets-numbers-violent-crime-drugs-addiction-treatment-homelessness-mayor-michael-hancock/> (accessed July 2, 2024).

RTD FasTracks (2014). Haunted Denver Union Station. [YouTube video; channel: RTD FasTracks]. *YouTube*. <https://www.youtube.com/watch?v=1uWEiTrG1E8>

Stevens, M. E. (1974). Union Station: National Register of Historic Places Inventory – Nomination Form. [Register No. 74000571]. *National Register of Historic Places*. <https://npgallery.nps.gov/GetAsset/47425798-482f-4ad0-8984-f3ae30a1b7dc>

Union Station (n.d.). Our History: A Century and a Half in the Making. *Denver Union Station*. <https://www.denverunionstation.com/about/our-history/>

Walden, S. (2020). Denver's Welcome Mizpah Arch. *Colorado Railroads*. <https://www.corailroads.com/2020/03/denvers-welcome-mizpah-arch.html>

9

The Ghost and the Teapot Dome Scandal: The Grant-Humphreys Mansion

The Grant-Humphreys Mansion, located at 770 Pennsylvania Street in Denver, is an elegant 18,000 square foot home boasting forty-two rooms including a theatre and a bowling alley. It's the kind of home fit for the most prominent of individuals and once did serve as a private residence housing some remarkable families, one of which ended up being involved in one of the most significant political scandals of its time. Now operated by History Colorado and used primarily as a wedding and event venue, the great neoclassical mansion is also home to its share of ghostly rumors and apparitions.

Figure 9.1. The Grant-Humphreys Mansion. Photo: Bryan Bonner.

The History

Every old house has a history. Even the smallest and most insignificant of low-income homes, if it stands long enough, is going to see its share of love and loss, triumph and tragedy. Often, those stories get lost to history and new residents may have no idea what occurred in their home years or decades before they moved in. When the home in question happens to be one of the most magnificent mansions in Denver, however, the history tends to be written down and remembered. In the case of the Grant-Humphreys Mansion, that's both for better and for worse because this historic property saw both triumphs and tragedies on the scale few of us could ever imagine.

The mansion's history begins with a man named James Benton Grant (1848-1911). Prior to the Civil War, Grant's family was wealthy enough to own a plantation in his home state of Alabama. However, after the Civil War, in which Grant enlisted and served in the Confederate Army, their fortune was largely lost, but a wealthy family member managed to pay for Grant's education in mineral engineering.[1] He came to Leadville, Colorado in 1877 during a silver rush and oversaw a smelting operation on behalf of out of state investors, but by 1883, after marrying Mary Matteson Goodell (herself a prominent member of Denver society, Daughter of the American Revolution, and philanthropist) in 1881, he relocated to Denver and reorganized the business in his own name as the Omaha and Grant Smelter.[2] The Grants had two sons. Lester Eames Grant was born on March 21, 1884, and James Benton Grant, Jr. was born on May 6, 1888.[3]

In the same year as his move to Denver, he became the third man overall and the first Democrat to serve as Governor of Colorado, remaining in office from 1883 to 1885, though he was largely uninterested in politics and declined to seek reelection.[4] Instead, he took a leadership position at the Denver National Bank and established a home in a (now-demolished) Quality Hill mansion. However, after spending about a decade there, he decided he wanted a better home and commissioned architects Theodore Boal and Frederick Harnois to build something to rescue him from what Boal called the "sickening sameness" of Denver's mansions of the time.[5] In that, they certainly succeeded. The house was and remains an elegant and unique structure in the Beaux-Arts neoclassical style that stands as a reminder, in a world blighted by too many near-identical "McMansions," that architecture is not only a craft but an *art*, and that one's home is not

1   Noel, T. J. (2011). Grant-Humphreys Mansion. *Colorado Encyclopedia.* <https://coloradoencyclopedia.org/article/grant-humphreys-mansion> (accessed July 8, 2024).
2   Goodstein, P. (1996). *The Ghosts of Denver: Capitol Hill.* Denver, CO: New Social Publications
3   Cannon, H. (1964). First Ladies of Colorado: Mary Goodell Grant. *Colorado Magazine* 4(1): 26-33.
4   Zimmer, A. (2019). Colorado Governors: James B. Grant. *Colorado Virtual Library.* <https://www.coloradovirtuallibrary.org/resource-sharing/state-pubs-blog/colorado-governors-james-b-grant/> (accessed July 8, 2024).
5   Goodstein, P. (1996). *The Ghosts of Denver: Capitol Hill.* Denver, CO: New Social Publications

merely a place to hang one's hat and store one's stuff but a thing to be enjoyed for its own sake. Especially, we suppose, if one happens to belong to one of the city's wealthiest families. But we think the same argument is true at every socio-economic stratum.

Reports of the Mansion's earliest history sometimes vary in their details but the broad strokes of the story are clear. The building was constructed for a cost of somewhere between $35,000 and $75,000 (equivalent to about $1.27 million to $2.74 million in 2024) depending on which records you believe--$35,000 according to a list of early Denver building permits[6], $40,000 according to local historian Phil Goodstein[7], up to $75,000 according to "Doctor Colorado" Tom Noel.[8] It was (and is) remarkable not only for its architectural style but for its amenities. It featured a theater, an indoor shooting range later converted into a bowling alley, a billiard room, and a ballroom (which itself would serve as an American Red Cross workroom during World War II).[9] One part of the ceiling in just one room contains an architectural feature that differs from the rest of the design. It was added to create an intentional flaw so as not to offend God by attempting to create anything perfect.

Regardless of the pricing details, the home was completed and ready for the Grants to move in in 1902 (some reports, though fewer in number, say 1903). Unfortunately, Mr. Grant would suffer a heart attack the very same year and would spend most of the rest of his life tending to his health and enjoying a sort of retirement (to the extent anyone of such wealth and ambition ever truly "retires") in the outdoors. Mr. Grant died of heart disease on November 1, 1911 in Excelsior Springs, Missouri, but is buried in Denver's Fairmount Cemetery where he was joined by his wife after her death in 1941.[10]

Mary Goodell Grant did not spend the entirety of the remainder of her life at the Mansion, though she did remain for several more years (and for their parts, Lester Eames Grant lived in the home from 1902 to 1908 and James Benton Grant, Jr. from 1902 to 1914). But by 1917, it was time for the widow to sell and she found eager buyers in Albert E. Humphreys (1860-1927) and his wife Alice Boyd Humphreys.[11]

Albert E. "A. E." Humphreys was born in Virginia in 1860 and made his for-

6   Brantigan, C. O. (1998). Denver building permits 1889 to 1905. *Denver Public Library*, Special Collections, C354.647888 B735de.

7   Goodstein, P. (1996). *The Ghosts of Denver: Capitol Hill*. Denver, CO: New Social Publications

8   Noel, T. J. (2011). Grant-Humphreys Mansion. *Colorado Encyclopedia*. <https://coloradoencyclopedia.org/article/grant-humphreys-mansion> (accessed July 8, 2024).

9   Fink, R. (1970). Grant-Humphreys Mansion: National Register of Historic Places Inventory – Nomination Form. [Register No. 70000160]. *National Register of Historic Places*. <https://s3.amazonaws.com/NARAprodstorage/lz/electronic-records/rg-079/NPS_CO/70000160.pdf> (accessed July 8, 2024).

10   Cannon, H. (1964). First Ladies of Colorado: Mary Goodell Grant. *Colorado Magazine* 4(1): 26-33.

11   Noel, T. J. (2011). Grant-Humphreys Mansion. *Colorado Encyclopedia*. <https://coloradoencyclopedia.org/article/grant-humphreys-mansion> (accessed July 8, 2024).

tune in the oil industry. He was known as "King of the Wildcatters" (a term referring to the drilling of "wildcat wells" or oil wells in areas not known to produce
oil) for his successes in that business in Wyoming, Oklahoma, and Texas, though
he also had successful lumber and mining ventures in Virgina, Minnesota, British
Columbia, and Colorado, where he moved with his wife Alice Boyd in 1989.

By all accounts Colonel Humphreys (a title he earned during militia service
early in his life and which stuck with him) was a devoutly religious man who
dedicated much of his wealth to religious and charitable service. In one story, he
prayed during an illness that he would donate even more of his wealth to charity
if he recovered and, after recovering, spent much of the remainder of his life trying to fulfill his promise.[12] Those donations often supported the foundation and
maintenance of Baptist schools, churches, and a variety of other philanthropic
missions.[13]

Unfortunately, despite his reputed financial scrupulosity and these various
philanthropic endeavors, Humphreys' life is most commonly remembered now
for his peripheral involvement in the most notorious political scandal of its day:
the Teapot Dome Scandal. Lest you think we're overstating the matter, most historians consider it the most significant scandal in American politics until it was
overshadowed by Watergate.

Entire books have been dedicated to the history of the scandal. If any reader is interested in learning all the intricacies of a complicated piece of political
history with numerous characters and many moving parts, we refer you to those
works to get a more complete picture (several such books are listed in the short
bibliography at the end of this chapter). Our treatment of the subject will be only
a brief sketch so you can understand the gist of the scandal itself and how Humphreys got involved. Writing about century-old political controversies, apparently,
is Rocky Mountain Paranormal's idea of "speaking truth to power."

The story begins with President William Howard Taft (1857-1930; President from 1909-1913), who had designated several locations as naval oil reserves
to ensure the United States Navy would have enough reserve fuel to continue
operations in times of unexpected conflict or during oil shortages. Three such
oil reserves were the Teapot Dome Oil Field in Natrona County, Wyoming, and
the Elk Hills and Buena Vista Oil Fields in Kern County, California. On May
31, 1921, President Warren G. Harding (1865-1923; President from 1921-1923)
issued Executive Order 3474 which transferred control of these fields from the
Department of the Navy to the Department of the Interior and authorized the
lease of these fields for drilling "under supervision of the President."[14]

---

12   Goodstein, P. (1996). *The Ghosts of Denver: Capitol Hill.* Denver, CO: New Social
Publications
13   Noel, T. J. (2011). Grant-Humphreys Mansion. *Colorado Encyclopedia.* <https://
coloradoencyclopedia.org/article/grant-humphreys-mansion> (accessed July 8, 2024).
14   Harding, W. G. (1921). Executive Order 3474-Transferring Naval Petroleum
Reserves in California and Wyoming, and Naval Shale Reserves in Colorado and Utah,
Under the Control of the Interior Secretary, Under Supervision of the President. *The
American Presidency Project.* <https://www.presidency.ucsb.edu/documents/executive-
order-3474-transferring-naval-petroleum-reserves-california-and-wyoming-and>

Secretary of the Interior Albert Bacon Fall (1861-1944) acted upon this newfound authority in 1922 to lease Teapot Dome oil production rights to Harry F. Sinclair of Mammoth Oil, a subsidiary of Sinclair Oil, and Elk Hills oil production rights to Edward L. Doheny of the Pan American Petroleum and Transport Company. So far so good. This just seems like ordinary business.

What turned this into a scandal is that Fall granted these leases without any kind of competitive bidding process, and the leases themselves were disproportionately favorable to the oil companies. Even still, under the Mineral Leasing Act of 1920, this was not criminal.[15] Though it was legal, it angered these companies' competitors who felt cut out of the deal. They in turn wrote their representatives in the legislature and Senator John B. Kendrick (1857-1933; United States Senator from Wyoming from 1917-1933) opened an investigation. The results of that investigation? The transactions themselves, while questionably ethical, were perfectly legal. However, it turned out they were the results of bribery, which was emphatically *not* legal. What were the bribes? The first was a no-interest loan from Doheny to Fall in the amount of $100,000 (equivalent to about $1.75 million in 2024), along with assorted "gifts" Fall received from both Doheny and Sinclair totaling some $400,000 (just over $7 million in 2024).[16] And despite Fall's best efforts to keep these bribes secret, investigators noticed his newfound affluence and eventually traced the money. He became the first cabinet-level official sentenced to prison as a result of his misconduct, though he was surprisingly the only individual convicted in the scandal. A legend holds that the term "fall guy," referring to an individual meant to take most or all of the blame for a conspiracy, originates from Albert Fall, but this turns out to be too cute to be true. Though the phrase's etymology is unknown and debated, it most likely is associated with "taking a fall" rather than any individual's name.

As for President Harding himself, there's no evidence that he was involved in (or even aware of) the bribery scandal, though his close association with the key players most assuredly damaged both his Presidency and his reputation. He once said to newspaperman William Allen White, "I have no trouble with my enemies. I can take care of my enemies all right. But my damn friends, my goddamned friends, White, they're the ones who keep me walking the floor nights!"[17] Historians still debate how much of his associates' bad behavior the President may or may not have been aware of. Regardless, President Harding died, aged 57, of a sudden heart attack on August 2, 1923, just over two years into his first term as President. Conspiracy theories alleging either murder or suicide continue even to this day. Though there's no evidence to support the conspiracy theories, it is at least plausible that stress from the impending scandals may have exacerbated any of the President's health conditions and led to his untimely demise.[18]

(accessed July 8, 2024).

15   Subsequent legislation has significantly tightened the regulations surrounding such business deals, but the details of mineral leasing law are beyond the scope of this book.

16   History.com editors (2017). Teapot Dome Scandal. *History.com*. <https://www.history.com/topics/roaring-twenties/teapot-dome-scandal> (accessed July 8, 2024).

17   *Ibid.*

18   Daugherty, G. (2023). Why President Warren G. Harding's Sudden Death Sparked

So that's Teapot Dome in a nutshell. How in the world did Humphreys get involved in all that? In a word: indirectly. Again omitting many of the details because it involves a series of complicated political scandals, Humphreys was very well connected with many of the players in the Teapot Dome scandal. And though he wasn't a part of the particular transactions scrutinized in the scandal, he had participated in some ethically questionable oil transactions of his own, involving some of the same players. Most notably, he sold 333,333,333 barrels of oil to a certain Continental Oil of Canada, actually a dummy corporation set up by Harry M. Blackmer, Robert W. Stewart, and, notably, Harry F. Sinclair. Humphreys sold the oil to these men through the shell corporation for $1.50 per barrel; they then sold it to their own companies for $1.75 per barrel, collectively earning themselves a substantial fortune in free money.[19] That money ultimately ended up in a fund used to support Harding's Presidential campaign, which in turn led to Presidential appointments favorable to the oil interests.

But that still wasn't Teapot Dome itself. Humphreys involvement in that was simply because he was friends with and did business with many of the key players. By 1926, the cat was out of the proverbial bag and Humphreys received a subpoena to testify in Fall's trial. Marking the only time one might actually be grateful for illness, his appearance was postponed because he had a serious case of pneumonia affecting both lungs. But by 1927, he'd recovered enough that his appearance at the trial was rescheduled.[20]

On May 8, 1927, the Humphreys were hosting an early Sunday dinner attended by several friends, family, and acquaintances. Shortly after finishing his meal, at about 4:00 p.m., Humphreys excused himself from the party and announced he was going to the den to clean his favorite hunting rifle in preparation for a hunting trip he planned to take before his mandated appearance at the trial. The other guests then heard the unmistakable sound of a gunshot which they determined originated from Humphreys' room. When they went to investigate, they found that he'd been shot in the head. Still alive but missing his lower jaw, Humphreys was rushed to the hospital where he died later that night.[21]

Reports vary on the details. The above is the most commonly-repeated version of the story. Others have it that rather than cleaning his rifle, Humphreys

Rumors of Murder and Suicide. *Smithsonian.* <https://www.smithsonianmag.com/history/why-president-warren-g-hardings-sudden-death-sparked-rumors-of-murder-and-suicide-180982626/> (accessed July 8, 2024).

Stolz, J. L. (2023). Presidential history: The puzzling death of Warren G. Harding. *Daily Press.* <https://www.dailypress.com/2023/06/07/presidential-history-the-puzzling-death-of-warren-g-harding/> (accessed July 8, 2024).

19   Noel, T. J. (2009). Noel: Where the ghost of A.E. still roams. *The Denver Post.* <https://www.denverpost.com/2009/10/22/noel-where-the-ghost-of-a-e-still-roams/> (accessed July 8, 2024).

McCartney, L. (2008). *The Teapot Dome Scandal: How Big Oil Bought the Harding White House and Tried to Steal the Country.* New York: Random House.

20   Goodstein, P. (1996). *The Ghosts of Denver: Capitol Hill.* Denver, CO: New Social Publications

21   *Ibid.*

was packing for his trip when the incident took place. And reports differ on whether the shooting took place in the library (just south of the dining room) or in the third floor bathroom on the north side of the building.

The shooting was officially deemed an accident but few people actually believe it. One conspiracy theory noted that Humphrey's personal nurse, J. J. Schasky, disappeared after the accident and suggested perhaps the nurse had really been a hired assassin to prevent Humphreys' testimony, though historian Phil Goodstein thinks (and we agree) Schasky's disappearance was either entirely coincidental or simply the result of someone wanting to get away from an increasingly fraught situation.[22]

More likely, the shooting was neither an accident nor an assassination but a suicide. Humphreys left a note discovered after the shooting which read, "Please, doctor, let me cash in."[23] Perhaps that's not the most explicit suicide note in the world—and does seem a bit out of character for someone as reputedly religious as Humphreys—but it does nevertheless seem to allude to suicide. Why would Humphreys take his own life? After all, he was only called to testify; he wasn't on the hook for the scandal himself. Goodstein thinks he suffered from a conflict of conscience—on the one hand, he was a devoutly religious man who could not bear the thought of lying under oath during his testimony; on the other hand, he similarly couldn't bear the thought of being forced to betray his friends' trust during the same testimony.[24] Regardless of the exact details of his motivation, it's almost unthinkable that an accomplished shooter (which Humphreys was) would violate the fundamental rules of firearm safety[25] so disastrously as to accidentally shoot himself in the head. Humphreys' manager, A. A. King, has the last word on the matter: "I know A. E. Humphreys took his life brooding over this affair to shield some men of affairs."[26]

Following the shooting, Alice Humphreys continued to live in the mansion with their two sons (both of whom became notable in the aviation industry and are listed in the Colorado Aviation Hall of Fame), Ira and Albert, Jr., who lived in

---

22   *Ibid.*

23   *Ibid.*

24   *Ibid.*

25   Beginning shooters are always taught four fundamental rules and required to train until they become habitual: assume every firearm is loaded unless you've checked it yourself (and even then, treat it as if it's loaded), never point a firearm at anything you're not willing to destroy, always be aware of your target and its surroundings (including everything behind the target), and always keep your finger off the trigger until you're ready to shoot. For Humphreys' shot to have been an accident, he (an experienced shooter) would have had to simultaneously violate all four rules. It's true that the four rules weren't expressed in those terms during Humphreys' life—they were formalized and popularized by Lieutenant Colonel Jeff Cooper (1920-2006) decades after Humphreys' death—but they are merely a popular expression of ideas that would have been quite familiar to anyone who spent time around firearms, as Humphreys did.

26   Quoted in: Noel, T. J. (2009). Noel: Where the ghost of A.E. still roams. *The Denver Post.* <https://www.denverpost.com/2009/10/22/noel-where-the-ghost-of-a-e-still-roams/> (accessed July 8, 2024).

the house until his sudden death in 1968. At that point, Ira took over the mansion and the family's affairs. Upon his own death in 1976, Ira bequeathed the mansion to the Colorado Historical Society (now History Colorado), which renovated the property and now operates it as a wedding and event center.[27]

The Grant-Humphreys Mansion was placed on the National Register of Historic Places in 1970[28] and was listed as a Denver Individual Landmark in 1976.[29] Over the years, each owner has put his, her, or their own touches on the building. Rooms originally designed for one function have been repurposed to fulfill other needs. Garages have been added, and repairs have been made. And yet, despite its changes throughout its more than a century of life, the Mansion still retains much of its original charm.

Paranormal Claims

Of course, in addition to the history and charm alluded to above, the Mansion also collects plenty of ghost stories. It's even been suggested that rather than an accident, suicide, or assassination, Humphreys himself may have been killed by a ghost.[30] But as far as we know, that's only ever been suggested in jest (we hope).

It would be a lot more plausible, we think, to suggest that the ghost of Albert E. Humphreys himself now haunts the Mansion, and indeed that is the most commonly told spook story attached to the property. Reports vary a bit in their details, but numerous people have repeated stories of Humphreys' ghost appearing throughout the property. In some versions of the story, he tries to tell witnesses of his innocence (or perhaps, in another variation, to finally give the testimony his death prevented). In some tellings of the tale, part of the apparition's head is missing.

No particular room seems to be a primary target of these hauntings. Perhaps the ghost feels at home in the entire house. Or, interpreting the stories through the lens of folkloric study, perhaps the ghost stories aren't attached to a particular room because the actual room in which Humphreys shot himself isn't

---

27   History Colorado (n.d.). Spaces at Grant-Humphreys Mansion: History. *History Colorado.* <https://www.historycolorado.org/grant-humphreys-mansion-spaces> (accessed July 8, 2024).

28   Fink, R. (1970). Grant-Humphreys Mansion: National Register of Historic Places Inventory – Nomination Form. [Register No. 70000160]. *National Register of Historic Places.* <https://s3.amazonaws.com/NARAprodstorage/lz/electronic-records/rg-079/NPS_CO/70000160.pdf> (accessed July 8, 2024).

29   City and County of Denver (2023). Denver Individual Landmarks. City and County of Denver. <https://www.denvergov.org/files/assets/public/v/4/community-planning-and-development/documents/landmark-preservation/individual_landmarks_list.pdf> (accessed June 18, 2024).

30   Goodstein, P. (1996). *The Ghosts of Denver: Capitol Hill.* Denver, CO: New Social Publications

known to historians.[31]

Some people believe that, in addition to the ghost of Humphreys, the Mansion is haunted by ghosts of several of his former employees. Often these stories are told with a tone to suggest these ghosts haunt the location because of their mistreatment by Humphreys during their lives, though we hasten to point out the historical record hasn't given us any reason to think Humphreys abused his staff. Perhaps he did, perhaps he didn't, but that's all pure speculation. Equally plausible, if one wants to view Humphreys in a more positive light, is that the ghosts remain because they're still loyal to their former employer.

Because the Mansion is now commonly used as a wedding venue, a room has been designated as a bridal dressing room. Several brides making use of the room have reported seeing the shape of a man on the balcony to the room's north. Others, while standing in front of the mirror to prepare themselves for the big event, have reported seeing the reflection of someone who wasn't really in the room.

Multiple guests and employees have reported hearing footsteps or muffled voices coming from unoccupied areas of the Mansion. Others have seen "shadows" wandering the rooms and halls.

There's a rumor that there may have been another shooting at the location, complete with stories that victim also haunts the Mansion.

Finally, another ghost hunting team claimed, on their investigation of the Mansion, to have recorded an EVP in the bowling alley of a little girl speaking French. That ghost's identity is not known.

Our Investigation

We were contacted by the Mansion's management to help facilitate a ghost hunt evening as a fundraiser. Though we usually prefer to conduct investigations alone rather than with members of the public along for the ride, we enthusiastically agreed to help with the event to whatever extent we could. Even if the investigation yielded contaminated data, we figured, it should be a fun evening and the fundraiser was for a good cause.

Before we got there, we did some background research to see if the historical record could tell us anything about the ghost stories. Largely we came up blank. There's a lot of history to learn about the Mansion, but nothing to shed any particular light on the paranormal claims. The rumors about another shooting at the Mansion seem to be just that. While it's certainly possible such an event took place, and even possible it could have precipitated a haunting, we were unable to find any record of the crime/accident.

As for the little French girl in the bowling alley, we came up with little there as well. So far (though we're always happy to keep looking) we haven't found anything that might shed light on the alleged ghost's identity. It's all but certain there have been little French girls in the Mansion throughout its history, especially since it's now used for rentals to the public. But we simply have no idea who she's

---

31   We almost said "the room in which Humphreys died," but that location actually is known: he died at the hospital shortly after the shooting.

supposed to be (or have been). We did check on the alleged EVP recording from the other ghost hunting team in which they claimed to have heard the French girl, but fluent French speakers we consulted on the matter couldn't make out any of the words.

For our own on-site investigation, we arrived at about 5:00 p.m. to tour the location and start setting up our equipment. It occurs to us that most people end their day at about the same time we start ours. Makes for interesting sleep schedules. Guests began appearing for the evening fundraiser while we were setting up and several of them decided to assist, which was of course very much appreciated. We're still not changing our minds about keeping most of our investigations to a small crew to minimize the risk of contaminating data, but we have to admit the extra sets of hands were nice.

As always when investigating at a large location, we couldn't rig up the entire Mansion for monitoring, so we selected several locations to watch throughout the evening. Based on the locations of the most recent (at the time) reports of ghostly activity, we chose to monitory the bowling alley, the theatre, and the bridal dressing room.

Figure 9.2. The Grant-Humphreys theatre. Photo: Bryan Bonner.

The bowling alley was equipped with two video cameras, one at the end of the room and one in the middle, both facing north, as well as a natural (DC) EMF meter in view of the cameras. The theatre received video cameras at the base of the stairs facing south and at the back of the room facing the stage. We also placed a microphone at the theatre entrance. For the bridal dressing room, we set up video cameras at the main door facing the bed, on the bed monitoring two foam blocks used as control objects, in the corner of the room facing the bathroom door and mirror, and at the bathroom door facing the bed and balcony window. We also put a DC EMF meter on the bed in view of one of the cameras, and a microphone in the middle of the room.

Everything was in place by about 6:30 p.m. and we began taking our hourly EMF, temperature, and seismic readings. Things progressed slowly. When we weren't engaged in silent monitoring, we spent some time answering questions of the guests who were there with us, but the ghosts didn't seem to be eager to manifest themselves. By about 10:00 p.m. the guests got tired and left us alone (except for Mansion employees) to finish our investigation under conditions closer to what we're accustomed to. We wrapped up and packed up our things at about 4:00 a.m.

What did we see? A whole lot of nothing in this case. Temperatures dropped by about two degrees Fahrenheit over the course of the evening, which was exactly what we expected as the night air began to chill. EMF readings—both the continuous ones under video monitoring and the hourly ones taken throughout the Mansion—were consistent throughout the evening. All readings were between 0 and 1.5 milligauss, which is in line with what we'd expect from such a property. We didn't find any unusual vibrations or other seismic readings, and we didn't capture anything on video. Audio was of some minor interest. Though we didn't record anything anomalous, we did notice the acoustics of the building are remarkable. It's possible to hear voices or footsteps from several rooms away. We're not going to go as far as to say the building's acoustics could account for all the ghostly sounds people have reported over the years, but it is something that needs to be taken into consideration when evaluating the ghost stories.

Once again, we find ourselves in the disappointing situation of having to report no findings on our investigation.[32] It's always a lot more fun when we actually experience something. Sometimes we can explain them naturally and that's fun from a scientific perspective; sometimes we can't and that's fun from a paranormal perspective. But when nothing happens, we have to just be comfortable admitting it and hope to try again another time.

In the case of the Grant-Humphreys Mansion, we certainly do hope to return for another investigation some day. It's a remarkable building in its own right, connected to some fascinating history, and home to enough ghost stories that our stubborn streak won't let us let it go. Maybe next time we'll find the evidence we're looking for, but until then we can at least enjoy the stories and the history for their own sake.

---

32   One of your authors also spent some additional time at the Grant-Humphreys Mansion over the course of a couple of years when he received Colorado Mathematics Awards which were then given at the Mansion. Despite best hopes to the contrary, nothing paranormal happened during those events either.

References & Further Reading

Bates, J. L. (1963). *The Origins of Teapot Dome: Progressives, Parties, and Petroleum*. Urbana, IL: University of Illinois Press

Brantigan, C. O. (1998). Denver building permits 1889 to 1905. *Denver Public Library*, Special Collections, C354.647888 B735de.

City and County of Denver (2023). Denver Individual Landmarks. City and County of Denver. <https://www.denvergov.org/files/assets/public/v/4/community-planning-and-development/documents/landmark-preservation/individual_landmarks_list.pdf>

Daugherty, G. (2023). Why President Warren G. Harding's Sudden Death Sparked Rumors of Murder and Suicide. *Smithsonian*. <https://www.smithsonianmag.com/history/why-president-warren-g-hardings-sudden-death-sparked-rumors-of-murder-and-suicide-180982626/>

Fink, R. (1970). Grant-Humphreys Mansion: National Register of Historic Places Inventory – Nomination Form. [Register No. 70000160]. *National Register of Historic Places*. Archived from the original at <https://s3.amazonaws.com/NARAprodstorage/lz/electronic-records/rg-079/NPS_CO/70000160.pdf>

Goodstein, P. (1996). *The Ghosts of Denver: Capitol Hill*. Denver, CO: New Social Publications

Harding, W. G. (1921). Executive Order 3474-Transferring Naval Petroleum Reserves in California and Wyoming, and Naval Shale Reserves in Colorado and Utah, Under the Control of the Interior Secretary, Under Supervision of the President. *The American Presidency Project*. <https://www.presidency.ucsb.edu/documents/executive-order-3474-transferring-naval-petroleum-reserves-california-and-wyoming-and>

History.com editors (2017). Teapot Dome Scandal. *History.com*. <https://www.history.com/topics/roaring-twenties/teapot-dome-scandal>

McCartney, L. (2008). *The Teapot Dome Scandal: How Big Oil Bought the Harding White House and Tried to Steal the Country*. New York: Random House

Noel, T. J. (2009). Noel: Where the ghost of A.E. still roams. *The Denver Post*. <https://www.denverpost.com/2009/10/22/noel-where-the-ghost-of-a-e-still-roams/>

Noel, T. J. (2011). Grant-Humphreys Mansion. *Colorado Encyclopedia*. <https://coloradoencyclopedia.org/article/grant-humphreys-mansion>

Noggle, B. (1965). *Teapot Dome: Oil and Politics in the 1920s*. New York: Norton.

Stolz, J. L. (2023). Presidential history: The puzzling death of Warren G. Harding. *Daily Press*. <https://www.dailypress.com/2023/06/07/presidential-history-the-puzzling-death-of-warren-g-harding/>

Werner, M. R. & Starr, J. (1959). *Teapot Dome*. New York: Viking Press.

Zimmer, A. (2019). Colorado Governors: James B. Grant. *Colorado Virtual Library*. <https://www.coloradovirtuallibrary.org/resource-sharing/state-pubs-blog/colorado-governors-james-b-grant/>

10

The Other Mommy: The Spruce Lodge

The Spruce Lodge is a small hotel in a small town with a big ghost story. It's located at 29431 US-160 in South Fork, a town of just over 500 permanent residents (more during the summer tourist season) approximately 125 miles southwest of Colorado Springs. It's a lovely guest lodge with a long history intricately connected with the surrounding town, and over the years it's been a hotel, a boarding house, a private home, a restaurant, and more.

The owners at the time of our investigation shared more ghost stories than one might expect from such a place. Many of them are the kinds of stories we commonly hear of supposedly haunted hotels, but there was one tale in particular that gave even us a case of the heebie jeebies.

Figure 10.1. The Spruce Lodge. Photo: Bryan Bonner.

The History

Once upon a time, the town of South Fork was a significant waystation in Colorado. You might not know it now—indeed, many Colorado residents have never even heard of South Fork and it currently boasts a population of, according to the 2020 Census, only 510 permanent residents.[1] But in the mid to late 1800s, what would later become South Fork was a major stagecoach stop for passengers traveling between Denver and other towns until the Denver and Rio Grande Railroad established rail lines in the area in 1883 forced the locals to find new sources of income.[2]

The solution was simple: the new railroad and the various mining operations in the area required timber. And the men to do it had already established a business in town. Brothers Charles A. Galbreath (b. 1875) and O. S. Galbreath, Jr. (1873-1934) had established a general store in South Fork in 1898 and post office in 1900, marking the beginnings of a business empire that would form the foundation of the entire South Fork economy for over a century. To fill the needs of the changing local economy, the Galbreath Brothers founded the Galbreath Tie and Timber Company in 1905, using the new rail lines to move their timber products throughout the United States.[3]

What resulted was a "company town" no matter what one means by the phrase. Workers employed by the Galbreaths' company were sometimes paid in notes redeemable only at the Galbreath general store or other town offices. In 1920, a fire, apparently started by a welding torch sparked too near a pile of sawdust, consumed the Mill and a fair portion of the town. According to legend, the ghost of a man named James Blankey haunts the old mill. Because of this haunting, it's said, some of the workers refused to return when the mill reopened.[4]

The Galbreaths responded by hiring gentlemen named Nichols and Adams to build a new store followed by a campground to attract tourists and a hotel. The hotel, originally the Galbreath Hotel and later renamed the Spruce Lodge, was built in 1926 and opened in 1927. In addition to eight bedrooms and three bathrooms, the original hotel featured a pool hall, restaurant, and barber shop. Some of the first employees at the hotel were former mill workers who were reportedly paid 9.5 to 10 cents per hour for their labor.

---

1   US Census Bureau (2020). South Fork town, Colorado. *United States Census Bureau.* <https://data.census.gov/profile/South_Fork_town,_ Colorado?g=160XX00US0872395> (accessed July 8, 2024).

2   Town of South Fork (n.d.). The History of South Fork. *Town of South Fork, Colorado.* <https://townofsouthfork.colorado.gov/about-south-fork/our-history-and-heritage> (accessed July 8, 2024).

3   Plucinski, J. (2008). Spruce Lodge: National Register of Historic Places Registration Form. [Register No. 08001009]. National Register of Historic Places. Archived from the original at <https://www.historycolorado.org/sites/default/files/media/documents/2019/5rn1043.pdf> (accessed July 8, 2024).

4   Because this isn't related to the Spruce Lodge itself, we've reported it as part of the history rather than the paranormal claims we'll address later. We're unable to investigate any paranormal claims at the mill as the property was sold and leveled.

An early guest at the hotel reported, "The rooms were so cold I may as well have slept outside. The walls were so thin a conversation could be held with the person in the next room without missing a word."[5] Despite the less-than-rave early review, the hotel remained successful. In 1938, it became one of the first buildings in town to receive electricity.

O. S. Galbreath was listed in directories as the manager of the property (as well as the lumber company) until his death in 1934. Sources differ on whether he died of a heart attack or drowned during a flash flood. Regardless, Charles Galbreath took over the enterprises and ran them until he sold the hotel in 1946 and eventually dissolved the Galbreath Tie and Timber Company.[6]

Additions to the property included a number of tourist cabins to the lodge's west in 1936. During renovations in 1948, two of the four were lost in a fire and a third damaged beyond repair. An additional Chalet was built on the site of the burned cabins in 1950—the remaining cabin still stands and is known as the Cook's Cabin as it was the residence of the lodge's cook for 30 years. In 2008, it was officially listed on the National Register of Historic Places.[7]

Throughout its life, the Spruce Lodge has passed through numerous hands, but has remained a constant reminder of the small town's history. It's still active today and continues to lodge individuals looking for a wide variety of natural and cultural centers within a short drive of its location or those simply looking for a rustic getaway with remarkable views. In its current iteration as of this writing, it also serves (since 2022) primarily as housing for adults participating in a religious "gap year" program intended to mentor young adults.[8]

Paranormal Claims

Most of the ghost stories we know of the Spruce Lodge were relayed to us by the owners of the property at the time of our investigation (not the current owners as of this writing), a husband and wife team with a three-year-old son.[9] Some of the claims are the kinds of things we expect to hear about any supposedly haunted hotel, but some of them are quite unique to the location and one in particular gave even jaded ghost afficionados like us a low-grade case of the wiggins.

---

5   Plucinski, J. (2008). Spruce Lodge: National Register of Historic Places Registration Form. [Register No. 08001009]. National Register of Historic Places. Archived from the original at <https://www.historycolorado.org/sites/default/files/media/documents/2019/5rn1043.pdf> (accessed July 8, 2024).

6   *Ibid.*

7   *Ibid.*

8   Spruce Lodge (n.d.). Historic Spruce Lodge: A rich history that goes back to the 1920s. *Spruce Lodge.* <https://www.sprucelodges.com/SpruceLodgeHistory.html> (accessed July 8, 2024).

9   For ease of reading, any time we refer to "the owners" without other explanation, we're referring to the owners as of our investigation, not the current or original owners of the property.

It all began before the owners even finalized the purchase of the property. When they were touring the Lodge and discussing the sale, they noticed the length of the main stairs and part of the main entrance were covered in fine powder. That seemed like unusual décor for a hotel. When asked, the former owners explained they were trying to record footprints of who—or indeed, what—ever was causing the footsteps they constantly heard in the middle of the night.[10] The former owners also admitted to their prospective buyers that they'd experienced lights turning on and off by themselves, doors opening and closing without apparent cause, and rocking chairs rocking even when no one sat in them.

Other past owners have reported (at least based on what the owners told us during our interviews) seeing a woman in a black dress in the second floor hallway. Another story relayed to one of the past owners claims a young girl (approximately seven years of age) told Lodge staff that she'd felt someone lie down next to her in bed the previous evening. Undeterred by these stories, the new owners bought the property anyway, and then they started experiencing unusual things for themselves.

The very first night the new owners were at the Lodge, the wife woke up feeling terrified in the middle of the night. A moment later, her husband also woke up, bolted out of bed, and shouted "what the hell was that?" Neither of them could remember any noises or anything else that might have woken them up.

A few months later, during a gathering in the hotel's restaurant, the topic of discussion turned to the various ghost stories they'd been told by the previous owners. When they began discussing the powder they'd seen on the floors, the second-floor smoke alarms suddenly went off. Some of the people went to investigate but just as they reached the top of the stairs, the alarms turned off. No one ever figured out either why they started or why they stopped.

Early the following year, one of the owners was in Room Six upstairs, trying to decide which runner looked best on the dresser. While she worked, the closet door behind her gently opened all the way and then slammed shut and locked itself. The owners told us they kept the door open but regularly found it shut and locked without explanation. On another occasion, when the same owner was on the phone at the front desk, she heard the unmistakable sound of someone descending the creaky main staircase, and several times heard people walking around on the second floor. Every time, no one else was staying at the Lodge when she heard these noises.

Midsummer that year, the family were busy playing in the back yard when their son (three years old at the time) looked up at the Lodge, roughly in the direction of Rooms One through Three, and asked "who is that girl?" Neither of his parents saw anyone and there were no guests or lodgers at the hotel at the time.

Early that fall, the mother of one of the owners was alone in the Lodge and also heard the sound of someone walking down the second floor hallway.

And then late that fall (as you can see, this was quite an eventful year for this family), the owner's son was sick one day and went to bed early. About an

---

10   Credit where it's due: that's a good simple way to try to narrow down the possible explanations for an anomalous experience.

hour later, his parents heard him talking to himself so they went in to see what was the matter. No one else was in the room, of course, so they asked who he was talking to. His reply sent shivers down his parents spines—and ours once we heard the story.

"The Other Mommy," he said.

When asked who the Other Mommy was, he said she looked "just like Grandma."

They asked what she'd been saying.

"She can't talk to me," the boy said.

For us, children carrying on conversations with invisible "Other Mommies" who can't talk is where we draw the line! Don't get us wrong; we love the story. We're horror nerds. But it's quite the unusual thing for one of the stories on an investigation to genuinely sound like the opening act of a world-class horror movie.

The final claim reported to us happened early the next year. While in bed, the owners heard a click sound and the fireplace at the end of the bed suddenly turned itself on. One of them got up to investigate and found the switch on the fireplace had been turned to the "on" position. That explained the clicking sound but posed an even more troubling question: who or what had flipped the switch?

Our Investigation

After reading the ghostly claims the owners told us, you can surely see why they asked us to check the place out. Indeed, those are the kinds of stories that get us excited as well so we were more than happy to volunteer our time. Because the Lodge was a bit farther from our base of operations than we'd prefer to commute daily, we arranged to stay at the Lodge for a two-night investigation. During our stay, we were the only people present other than the owners and their child.

Following our customary tour of the property, we established monitoring throughout the building. We put two video cameras and a microphone in the kitchen along with a foam ball to be used as a control object. The dining room also got two video cameras and a microphone in addition to several foam balls (placed within the cameras' view). On the main staircase, we placed a foam ball and a toy dinosaur within view of a video camera looking up the stairs. Then on the second floor hallway we placed several more foam balls and a DC EMF meter, all within view of two video cameras at opposite ends of the hall. Rooms One, Three, and Seven as well as the child's bedroom in the basement each got a single video camera and a single ball as a control object. The owners' master bedroom, also in the basement, received two video cameras and a ball as a control object. If anything happened in this place, we were ready to catch it.

Most of the investigation was uneventful, though. Our regular temperature and EMF readings didn't reveal anything out of the ordinary. Temperature changes were within the range expected as the outside temperature shifted throughout the day and night, and EMF readings ranged from 0 to 1 milligauss, well within normal expectations. We did detect slightly elevated EMF levels in Room One, but quickly determined these readings were due to a nightstand clock/radio near the head of the bed.

Figure 10.2. The haunted ball in the kitchen. Photo: RMP archives.

Just when we were about to give up, we got to record one notable event. On the second evening at about 10:30 p.m., the foam ball control object in the kitchen—which had been placed on the prep table—rolled about four inches and abruptly stopped. If you've been reading this book in order and think that's remarkably similar to what happened at the Bullock Hotel in Chapter 4, you're not wrong. And just as we did in that case, we spent the rest of the evening checking for drafts, vibrations in the flooring, or any other possible rational explanation for what we just witnessed. We never figured it out.

Either ghosts must really like rolling balls just a few inches when we're around or we have the weirdest luck in the world. Intriguing as these events are, of course, they fall short of the kind of "proof positive" that would allow us to unreservedly declare a location haunted. What they do, instead, is to provide us the kind of tantalizing hints that make our work exciting and keep us moving on to the next investigation. From time to time when we find ourselves in a rut in which all our investigations seem to be coming to nothing, it's the memory of these little pieces of weird, inconclusive though they are, that motivate us to keep up the search.

And so we temporarily close the book on our Spruce Lodge investigation having to leave the case, as with so many others, unsolved.

References & Further Reading

Plucinski, J. (2008). Spruce Lodge: National Register of Historic Places Registration Form. [Register No. 08001009]. National Register of Historic Places. Archived from the original at <https://www.historycolorado.org/sites/default/files/media/documents/2019/5rn1043.pdf>

Spruce Lodge (n.d.). Historic Spruce Lodge: A rich history that goes back to the 1920s. *Spruce Lodge.* <https://www.sprucelodges.com/SpruceLodgeHistory.html>

Town of South Fork (n.d.). The History of South Fork. *Town of South Fork, Colorado.* <https://townofsouthfork.colorado.gov/about-south-fork/our-history-and-heritage>

US Census Bureau (2020). South Fork town, Colorado. *United States Census Bureau.* <https://data.census.gov/profile/South_Fork_town,_Colorado?g=160XX00US0872395>

11

An Investigation for the Radio: Heritage Square Music Hall

This entire chapter is a ghost story. The Heritage Square Music Hall (originally the Heritage Square Opera House) was part of the Heritage Square amusement park and shopping center in Golden, Colorado. Unfortunately the "Storybook Victorian" entertainment venue closed its doors in 2013, followed by the closure of the entire park in 2018. The entire area is now an abandoned ghost town worthy of any horror story. Ghosts were said to haunt the music hall even when it was open, though, and we had the opportunity to conduct an overnight investigation as part of a Halloween radio program.

Figure 11.1. Heritage Square Music Hall. Photo: Bryan Bonner.

The History

This one is going to be brief but important. Though we don't know too many fascinating stories about the history of either the Music Hall itself or Heritage Square in general, we think it's worth preserving what little we do know about it. Few probably knew the history of the park when they visited, but for many people who grew up in the Denver area, those annual visits to Heritage Square were an integral part of their childhood. The park's closure was recent enough that plenty of us still remember it, but as we all age and our children take our places in society, the memory of this remarkable little theme park risks being as abandoned as the grounds themselves.

Theme parks are nothing new. They trace their lineage (at least) to Ancient Roman centers of amusement, through the "pleasure gardens" of the 18th and 19th centuries to the industrialized amusement parks of the late 19th and early 20th centuries (think of the Chicago World's Fair); however, it was the arrival of Disneyland in 1955 that established the modern model of a centrally planned and controlled theme park meant to provide families with safer entertainment than that afforded by the stereotypical carnies.[1]

One of the main architects of Disneyland was one C. V. "Woody" Wood. After a falling out with the Disney company, Wood would go on to work on other theme parks, billing himself as the "master planner" of Disneyland until litigation forced him to stop using that moniker.[2] Though Wood has been largely forgotten by chroniclers of Disneyland history, his work was known to Colorado businessmen Walter Francis Cobb and John Calvin Sutton who hired him to establish the park that would become Heritage Square—then known as Magic Mountain—between 1957 and 1959. Wood brought with him a team of established park designers and created the quintessential example of "storybook Victorian" design, a style clearly inspired by actual Victorian architecture but intended to give visitors a sense that they'd stepped into the pages of a storybook rather than a history museum or a time machine. In this case, the design was also infused with influences drawn from Colorado history.

Unfortunately, despite their grand vision for the park, it never quite found its footing. Though he was not entirely to blame for the park's plague of financial woes, Cobb stepped down as president shortly before Magic Mountain was forced into foreclosure. Both Cobb and William Zeckendorf (developer of much of New York's skyline) attempted to save the property, but to no avail. It closed its doors, seemingly for good, in 1960.[3] Many of its rides and attractions were auctioned off to other parks and the grounds themselves stood as a sort of ghost town and reminder of business failures.

---

1   Park Database (n.d.). A History of Theme Parks. *The Park Database.* <https://www.theparkdb.com/blog/history-of-theme-parks> (accessed July 9, 2024).

2   Schultz, J. (n.d.). Construction Men. *Magic Kingdom Chronicles.* Archived from the original at <https://archive.ph/20130129030239/http://www.mouseplanet.com/jason/006.htm> (accessed July 9, 2024).

3   Golden Landmarks (n.d.). The Rest of the Story. *Golden Landmarks.* <https://goldenlandmarks.com/the-rest-of-the-story/> (accessed July 9, 2024).

Resurrections don't just happen in religions and zombie stories, though. After more than a decade of disuse, the property was purchased by the Woodmoor Corporation, who set about reestablishing the park as Heritage Square. Rather than the theme park originally envisioned as Magic Mountain, they created a combination theme park and shopping village which opened its gates—which would not charge an admission fee, profiting instead from sales of goods or tickets for the attractions—in 1971.[4]

Locals probably remember Heritage Square best for its famous "Alpine Slide," a down-mountain slide so large it could be seen from the highway. But if you actually exited the highway and entered the park, you were sure to have a day of fun with rides including a Tilt-A-Whirl, swan boats, swings, trains, a Ferris wheel, and more. They were joined by shopping venues, restaurants, arcade games, public art, and a variety of entertainment venues.[5]

Among those was the Heritage Square Music Hall. It opened originally as the Heritage Square Opera House[6] in 1973 under the management of G. William "Bill" Oakley, Jr., a producer/director/actor best remembered now as a pioneer of melodrama who revived early American theatrical styles as a sort of nostalgic theatre.[7] In 1988, it was taken over by actor T. J. Mullin and rebranded the Heritage Square Music Hall. Mullin, along with Connie Helsley, kept the theatre running and mounting productions until its closure in 2013.[8] Many of us who were around during its operation remember it as a favorite dinner theatre venue.

Blame television, movies, video games, the Internet. Blame who- or whatever you want. Unfortunately, despite best efforts of many of us, interest in cultural events like live theatre seems to be fading away these days. The Heritage Square Music Hall was the first piece of the park to close its doors, citing declining ticket sales; nevertheless, despite attempts by mining company Martin Marietta, landlord of the park and owner of the adjacent quarry, to buy out the facility—and, according to a Colorado Public Radio report, to use their trademark of the name "Heritage Square" to force the park to use a different name—the park made a valiant attempt to continue operations amidst declining attendance and legal battles.[9] In 2018, the park's owners reached a settlement with landlord Martin Marietta Materials and permanently closed on June 30.[10]

---

4   *Ibid.*

5   Wilde, N. (2022). Popular Colorado Amusement Park Closed and Sadly Sold Off. *95 Rock.* <https://95rockfm.com/abandoned-colorado-amusement-park/> (accessed July 9, 2024).

6   And while we're on the subject of closed, forgotten, or endangered pieces of our history: don't forget to support your local opera company.

7   Colorado Theater History (n.d.). Heritage Square Music Hall. *Colorado Theater History.* <https://coloradotheatrehistory.com/heritage-square-music-hall/> (accessed July 9, 2024).

8   *Ibid.*

9   Awad, A. M. (2017). Heritage Square Is No More, But The Amusement Park Still Lives (For Now). *CPR News.* <https://www.cpr.org/2017/08/21/heritage-square-is-no-more-but-the-amusement-park-still-lives-for-now> (accessed July 9, 2024).

10   Kesting, A. (2018). Heritage Square closes for good. *9News.* <https://

The following year, Jefferson County and Martin Marietta proposed a land exchange which would see the Heritage Square grounds become a public recreation area and give the mining company opportunity to expand one of its existing mines into new territory.[11] The land exchange was completed between Jefferson County Open Space and Martin Marietta Materials in 2023, with the former planning to sell the Heritage Square property for zoned uses "based on proposals that best fit the site and surrounding areas, keeping historic celebration in mind."[12]

But for now, most of the buildings and attractions have been removed and the property remains a sort of ghost in and of itself.

Paranormal Claims

Though the entire Heritage Square grounds can be considered a metaphorical ghost, the Music Hall was, before it was torn down, supposed to be home to some literal ghosts of its own. Despite its haunted reputation, though, relatively few of those ghost stories have been popularized in the paranormal literature, so there's not a whole lot for us to share.

Fundamentally, the ghost stories come down to variations on three claims. First, a ghostly woman was sometimes seen in the balcony seating. Second, staff reported sometimes seeing items moving by themselves in the backstage area. Finally, various ghostly apparitions have been seen on stage after hours.

On that last point, though we know little of the history of any particular ghosts meant to haunt the Heritage Square Music Hall, we can say at least the story tracks with what we're familiar with from ghostly lore of other theatres.

Theatre people tend to be a superstitious lot. Many old stages, opera houses, music halls, and other forms of theatre have their own ghost stories. Sometimes they're tales of actors who died during rehearsals or performances. Other times they're supposed to be the ghosts of theatre lovers who just want to keep coming back to enjoy another show. Non-ghostly superstitions also abound. For instance, theatre people will often chastise anyone who dares to use the name of the play *Macbeth* instead of euphemistically calling it "the Scottish Play," particularly if one utters the cursed name in an actual theatre.[13]

One of our favorite theatrical traditions/superstitions, though, is that of the "ghost light." The tradition is simple: when everyone leaves the theatre at the end of an evening, a light, often placed center stage, is left on. It may have come from a practical consideration, making sure the theatre had some light should anyone need to come back during the evening or so the opening crew wouldn't

www.9news.com/article/news/local/heritage-square-to-close-for-good-on-saturday/73-567355459> (accessed July 9, 2024).

11   Sylte, A. (2019). Land that was once Heritage Square could become open space. *9News*. <https://www.9news.com/article/news/land-that-was-once-heritage-square-could-become-open-space> (accessed July 9, 2024).

12   Jefferson County Open Space (n.d.). Heritage Square Exchange. *Jefferson County Open Space*. <https://www.jeffco.us/3810/Heritage-Square-Exchange> (accessed July 9, 2024).

13   We've done it many times, and the curse hasn't done anything terrible to us yet.

trip on anything the next morning. But a superstitious connotation provides an even better reason: the light is left on for the benefit of any ghosts in the theatre. Whether one believes in ghosts literally or only likes the ghost stories for literary or metaphorical value, we think that's a great tradition. It's a nod to all the theatre people—actors, costumers, directors, technical crew, and more—who have "trod the boards" before one's own time on the stage. It reminds us of our history.

And, of course, if there actually were ghostly apparitions wandering the stage of the Heritage Square Music Hall after hours, we're sure they appreciated the courtesy.[14]

Our Investigation

Just as our catalog of ghost stories from this venue was brief, so will be our description of our investigation. It was a non-traditional investigation, conducted as much for show as for actual science, though even under those circumstances we still do our best to remain objective and to collect the highest quality data possible.

In this case, we were invited to participate in the Peter Boyles Halloween Special radio program for the 630 KHOW station. The investigation would begin the night before Halloween, extending into the morning of the holiday itself, and would be open to as many members of the public as the facility could manage for such an event. About twenty-five people showed up in addition to ourselves, other paranormal seekers and enthusiasts, venue staff, and radio crew.

Beginning the night before Halloween, we set up our usual array of monitoring equipment throughout the theatre, with particular emphases on the stage, the balcony, and the backstage area. And then we did our best to monitor silently. Initially this was difficult as there were so many people around. Other ghost hunters and psychics were doing their thing. Members of the public were following around and asking questions. It was a fun evening, but we didn't collect a lot of quality data.

Shortly after the evening broadcast, though, most of the other people who'd agreed to stay overnight got tired and went to sleep, leaving only ourselves and the other ghost hunters around. Still not the best monitoring conditions but at least quieter than earlier in the evening.

Not much happened during the evening, but we did catch one unusual event. At one point, directly in front of one of our investigators' eyes (and witnessed by several others), a door slowly opened as if someone was about to walk through, but no one was there. The door then shut itself.

This left us with some conflicting thoughts. On the one hand, we had already checked the doors in the Music Hall and we knew that door was securely shut. Yet on the other hand, this was also an evening of high winds outside the theatre—indeed, even from inside, we could often hear it howling. So what are we to make of this? Did a ghost announce his or her presence by opening and closing a door directly in front of us? Or was the door not quite as secure as we

---

14   We prefer not to think of the implications of the theatre's closing and destruction for any ghosts who might have been present.

thought and a gust of wind outside pushed a draft through the building in just the right place? Honestly, we have no idea. As much as we'd like to think it was a ghost, we have to admit the wind seems a far simpler explanation. But given the conditions and time constraints on this investigation, we never got a conclusive answer.

Figure 11.2. The stage, ready for our radio interview. Photo: Bryan Bonner.

After the investigation wrapped up, we were able to sit on stage with Peter Boyles for a live interview during his morning radio program. We might not have found the definitive proof of the paranormal we always hope to find on an investigation, but this did give us the opportunity for an extended chat with the radio host and his various call-in guests on topics ranging from philosophical musings on the nature of the soul and the meaning of life to individuals' own ghost stories to the loss of some of our favorite Halloween traditions of yesteryear to classic horror movies and the histories thereof. Anyone who knows anything about us is already aware those are among our favorite topics of conversation, so we have to say despite the brevity of our experiences there and lack of conclusions, it was a great experience for us.

Cases that end with such inconclusive findings are usually left open in our case files. Unfortunately the loss of the Heritage Square Music Hall means we have to permanently close this case without resolution.

References & Further Reading

Awad, A. M. (2017). Heritage Square Is No More, But The Amusement Park Still Lives (For Now). *CPR News.* <https://www.cpr.org/2017/08/21/heritage-square-is-no-more-but-the-amusement-park-still-lives-for-now>

Colorado Theater History (n.d.). Heritage Square Music Hall. *Colorado Theater History.* <https://coloradotheatrehistory.com/heritage-square-music-hall/>

Golden Landmarks (n.d.). The Rest of the Story. *Golden Landmarks.* <https://goldenlandmarks.com/the-rest-of-the-story/>

Jefferson County Open Space (n.d.). Heritage Square Exchange. *Jefferson County Open Space*. <https://www.jeffco.us/3810/Heritage-Square-Exchange>

Kesting, A. (2018). Heritage Square closes for good. *9News*. <https://www.9news.com/article/news/local/heritage-square-to-close-for-good-on-saturday/73-567355459>

Park Database (n.d.). A History of Theme Parks. *The Park Database*. <https://www.theparkdb.com/blog/history-of-theme-parks>

Schultz, J. (n.d.). Construction Men. *Magic Kingdom Chronicles*. Archived from the original at <https://archive.ph/20130129030239/http://www.mouseplanet.com/jason/006.htm>

Sylte, A. (2019). Land that was once Heritage Square could become open space. *9News*. <https://www.9news.com/article/news/land-that-was-once-heritage-square-could-become-open-space>

Wilde, N. (2022). Popular Colorado Amusement Park Closed and Sadly Sold Off. *95 Rock*. <https://95rockfm.com/abandoned-colorado-amusement-park/>

12

They Are All Around Us: Dunafon Castle

What kind of paranormal investigators would we be if we hadn't spent some quality time in a haunted castle? Fortunately, Colorado has one of its very own. The Dunafon Castle, located at 24020 CO-74 in Idledale, Colorado, about three and a half miles west of Red Rocks Park and Amphitheatre (see Volume 1, Chapter 11 and forthcoming volumes for more on Red Rocks), currently operates as a majestic events venue on 17 acres of property offering clients a fairytale wedding with a portion of proceeds going to a charitable cause.

As for ghosts, plenty of individuals who've spent time at the Castle have stories to tell. We were therefore thrilled to have the opportunity to explore and investigate the property in conjunction with another Halloween radio program.

Figure 12.1. Dunafon Castle. Photo: Bryan Bonner.

The History

Dunafon Castle's history begins in 1929 with Marcus and Muriel Wright. A water engineer by trade, Mr. Wright designed the Castle himself and began construction in 1929 following the tradition of 13th century Celtic castles including turrets, handmade light fixtures, tunnels, thick stone walls, and arched windows. Yet despite its nostalgic style and appearance, Mr. Wright included features well ahead of their time. Electricity wasn't yet available in the area, so Wright used his training as an engineer to provide power for his castle by means of a water wheel. He also made it wheelchair accessible fifty years before such became a regulation mandated by law.[1]

Construction was a long process, but Wright was able to find plenty of help. The stock market crash of 1929 left many men in need of work and Wright was happy to hire them on. Rock for the building itself was quarried from the Lair O' the Bear Park, which remains a public park in Idledale offering spectacular views, challenging trails for hikers, and access to the nearby Bear Creek for fishermen.[2] Today, the Castle is only about a four minute drive along Bear Creek Road from the Park by car. However, when construction began, workers had to haul the stones up the mountain by wheelbarrow. Soon, Wright realized he needed a better plan, and installed his own narrow-gauge railway to move the rocks to the construction site.[3] And he knew his engineering well. He used a mining train driven on all wheels, reasoning that the heavy loads would thus provide greater traction to pull the train over the mountain. During the construction process, Wright's team installed the track themselves, including a bridge and a house to store the train when not in use.

Construction was completed in 1941. The finished castle came in at just under 9,000 square feet and boasted five bedrooms, one full bathroom, three three-quarters bathrooms, three fireplaces, and one wood stove. It fit the true stereotype of a castle, complete with turrets, a moat, a bridge, and beautiful stone walls. The rest of the property included a four-car garage (itself equipped with a fireplace for heat), and a dam containing some 10,000 gallons of water in underground storage from a Bear Creek diversion to Wright's twenty-two-foot water wheel connected to a hydroelectric generator which supplied electricity and heat to the Castle and to the train. The land on which it sits includes no fewer than four ponds which are also used in the Castle's power and irrigation system.

Unfortunately, Mr. Wright lost his battle with cancer in 1965 and his magnificent Castle was placed into a trust. But without a permanent resident or active caretaker, it began to fall into disrepair until it was purchased in either 1969 or 1970 (reports vary) by William and Tamsin Barnes who moved in with their four daughters, Paula, Tamara, Lisa, and Elizabeth. William, born May 16, 1936,

---

1   Dunafon Castle (n.d.). About Us: A Rich History of the Castle. *Dunafon Castle*. <https://dunafon-castle.com/about-us> (accessed July 10, 2024).

2   Jefferson County Open Space (n.d.). Lair o' the Bear Park. *Jefferson County Open Space*. <https://www.jeffco.us/1254/Lair-o-the-Bear-Park> (accessed July 10, 2024).

3   Dunafon Castle (n.d.). About Us: A Rich History of the Castle. *Dunafon Castle*. <https://dunafon-castle.com/about-us> (accessed July 10, 2024).

was a computer engineer. He married Tamsin, born July 18, 1934, in 1960, who was a dedicated believer in animal rights. Both were members of the Young Republicans.[4] After their marriage, Tamsin began working as the sales manager for Autotrol, the computer mapping company William founded before eventually moving on to become the first female president of the American Society for Photogrammetry and Remote Sensing, a professional organization dedicated to mapping sciences.[5]

Figure 12.2. A Dunafon Castle stairway. Photo: Bryan Bonner.

Once they moved in, the Barnes family got the Castle back into top condition and started making some renovations of their own. They converted the four-car garage into an entertainment room complete with a kitchen and gambling equipment. There was even a claim that the Castle operated as an illegal gambling hall and brothel between Wright's death and the Barnes' purchase of the property.[6] We think the rumor may have started as a result of the Barnes' addition of their gaming room, but it's important to note both that we have never found any evidence to substantiate the rumor at all and, perhaps even more critically, the rumor never implicated any of the Barneses even if we assume there's some truth to it. What is undoubtedly true is that the Barnes family were engaged in a long-term project to rebuild, renovate, and modernize the Castle.

4   Denver Post (1999). Obituary: Tamsin Barnes, William Barnes, Paula Barnes: Flight 990 victims. *The Denver Post.* <https://extras.denverpost.com/news/obits/barn1118.htm> (accessed July 10, 2024).

5   ASPRS (n.d.). In Memoriam Archive: Tamsin G. Barnes 1934-1999. *American Society for Photogrammetry and Remote Sensing.* <https://www.asprs.org/a/news/archive/memoriam.html#Barnes> (accessed July 10, 2024).

6   Miller, D. (2013). Lair O' the Bear and Dunafon Castle. *Impression Evergreen.* <https://www.impressionevergreen.com/2013/02/lair-o-bear-and-dunafon-castle.html> (accessed July 10, 2024).

Whatever their future plans for the Castle may have been (some have suggested these included installation of a pool on the terrace), they were cut short by tragedy in 1999 when William, Tamsin, and daughter Paula were killed in the early hours of Halloween of 1999 in the crash of EgyptAir Flight 990.[7] Avid travelers, they were part of a group of fifty-four people setting out on a fourteen-day tour of the Nile.[8]

As with any deadly aviation accident, a full understanding of this crash requires the reader to go down a deep rabbit hole of news articles, investigative reports, and, in this case, international politics. But to summarize briefly, EgyptAir 990 was a flight on October 31, 1999 from Los Angeles to Cairo with a stop in New York carrying a total of 217 people (203 passengers and 14 crew) from seven countries. The Boeing 767 departed John F. Kennedy International Airport on schedule at about 1:20 a.m. and crashed in international waters in the Atlantic Ocean about 60 miles south of Nantucket at about 1:52 a.m. (EST).[9]

Because the crash took place in international waters, both the American and Egyptian authorities investigated the accident, in accordance with Internation Civil Aviation Organization rules. However, because the Egyptian Civil Aviation Agency (ECAA) lacked the resources of the American government, they allowed the American National Transportation Safety Board (NTSB) to take the lead in the investigation. The Americans quickly began to suspect it was an intentional criminal act rather than a mere accident. Their suspicions centered on the co-pilot Gamil Al-Batouti who, upon activating the autopilot which sent the plane into a dive, uttered the phrase "Tawakilt ala Allah," which translates to "I put my faith in God." Just two weeks after the accident, the NTSB suggested handing the investigation over to the FBI as a potential criminal matter, but the Egyptians, outraged by the accusation, exerted enough political muscle[10] to keep the investigation within the NTSB.[11]

At the conclusion of their investigation, the NTSB found that the crash was caused by the co-pilots faulty flight control inputs rather than any environmental or mechanical failures; though much attention was devoted to the co-pilot's mental state and potential motivations, they did not reach any conclusions regarding

---

7   Denver Post (1999). Obituary: Tamsin Barnes, William Barnes, Paula Barnes: Flight 990 victims. *The Denver Post.* <https://extras.denverpost.com/news/obits/barn1118.htm> (accessed July 10, 2024).

8   ASPRS (n.d.). In Memoriam Archive: Tamsin G. Barnes 1934-1999. *American Society for Photogrammetry and Remote Sensing.* <https://www.asprs.org/a/news/archive/memoriam.html#Barnes> (accessed July 10, 2024).

9   NTSB (2002). Aircraft Accident Brief: EgyptAir Flight 990, Boeing 767-366ER, SU-GAP, 60 Miles South of Nantucket, Massachusetts, October 31, 1999. *National Transportation Safety Board.* <https://www.ntsb.gov/investigations/AccidentReports/Reports/AAB0201.pdf> (accessed July 10, 2024).

10   This was during a time when President Clinton required Egyptian support for his ultimately doomed attempts at Arab-Israeli peace.

11   Borger, J. & Dawoud, K. (2000). Wings and a prayer: EgyptAir flight 990 crash: special report. *The Guardian.* <https://www.theguardian.com/world/2000/may/08/egyptaircrash.usa> (accessed July 10, 2024).

a reason—intentional or accidental—for his actions.[12] For their part, the ECAA, despite having initially ceded authority over the investigation to the Americans, refused to accept this result and issued their own conclusions: "The Relief First Officer (RFO) did not deliberately dive the airplane into the ocean. Nowhere in the 1665 pages of the NTSB's docket or in the 18 months of investigative effort is there any evidence to support the so-called 'deliberate act theory.' ...There is evidence pointing to a mechanical defect in the elevator control system of the accident. ...Although this evidence, combined with certain data from the Flight Data Recorder (FDR), points to a mechanical cause for the accident, reaching a definitive conclusion at this point is not possible...."[13]

So who was correct? One could certainly argue, in favor of the American conclusions, that the NTSB had far greater resources than the ECAA and therefore their report might be more trustworthy. But of course the two boards, despite their different governments and political loyalties, largely collaborated on the investigation (even if it was sometimes a fraught relationship). And they operated from the same pool of evidence, at least in drawing their published conclusions (one assumes both government agencies keep their various secrets, but their published documents are based on mutually-available evidence). One could also argue that, since the co-pilot in question was an Egyptian national, the ECAA had a vested interest in drawing their conclusions to save face. But one could equally argue that the NTSB might have been unduly suspicious of the co-pilot due to political considerations or motivated to blame the deceased co-pilot rather than the American airplane manufacturer. Media speculation ran wild in both Western and Egyptian media, ranging from sober reporting on the investigation's progress to outlandish conspiracy theories.

Though we tend to find the NTSB report more credible, it remains an open question. The crash does, however, have an even darker legacy. Without making any claims of our own regarding its cause or the motivations of any of the individuals involved, a stronger case can be made that none other than Osama bin Laden believed co-pilot Al-Batouti crashed the plane deliberately and drew inspiration from the incident for his own terrorist attack against the United States on September 11, 2001.[14]

For our purposes here, the important thing is simply to remember the 217 passengers and crew, none of whom survived the crash. A memorial to the victims is located in Newport, Rhode Island's Island Cemetery.

---

12   NTSB (2002). Aircraft Accident Brief: EgyptAir Flight 990, Boeing 767-366ER, SU-GAP, 60 Miles South of Nantucket, Massachusetts, October 31, 1999. *National Transportation Safety Board.* <https://www.ntsb.gov/investigations/AccidentReports/Reports/AAB0201.pdf> (accessed July 10, 2024).

13   ECAA (2001). Report of Investigation of Accident: EgyptAir Flight 990, October 31, 1999, Boeing 767-300ER SU-GAP, Atlantic Ocean-60 Miles Southeast of Nantucket Island. *Egyptian Civil Aviation Authority.* Published courtesy of NTSB and archived from the original at <https://web.archive.org/web/20110622102818/http://www.ntsb.gov/events/ea990/docket/ecaa_report.pdf> (accessed July 10, 2024).

14   Lahoud, N. (2022). *The Bin Laden Papers: How the Abbottabad Raid Revealed the Truth About al-Quaea, Its Leader, and His Family.* New Haven, CT: Yale University Press.

Returning to the Barnes family, they were survived by three daughters to William and Tamsin (sisters to Paula): Tamara Lewis, Lisa Berry, and Elizabeth Bergman; William Barnes was also survived by three brothers, a sister, and a step-mother.[15] William and Tamsin Barnes, through no fault of their own, also have the dubious legacy of their names being used in an early "Nigerian email" scam.[16] But their real legacies are their family, their work in both business and their community, and their improvements to the Dunafon Castle.

Unfortunately, it took some time before the Castle was again occupied. Following the Barnes' deaths, it passed into a trust and remained vacant until it was discovered in 2004 by its current owners, who purchased it the same year. Though it had again fallen into some disrepair and the grounds were overgrown, these individuals saw its potential and set about restoring it to its original glory. Their work was aided living members of the Wright and Barnes families as well as more than 200 men from the "Step 13" program, a charitable organization that provides transitional living assistance for homeless men trying to give up addictions and put their lives back in order.[17]

Though the renovations were still in progress at the time of our investigation, Dunafon Castle is now open for business and has become a much sought-after wedding venue. A portion of their booking fees still support Step 13, and many of their employees have come out of the same program.[18]

Paranormal Claims

Ghost stories of the Dunafon Castle came to us courtesy both of the current caretaker (at the time of our investigation) as well as the surviving Barnes children who spoke on the radio about their experiences just prior to our inves-

---

15   Denver Post (1999). Obituary: Tamsin Barnes, William Barnes, Paula Barnes: Flight 990 victims. *The Denver Post*. <https://extras.denverpost.com/news/obits/barn1118.htm> (accessed July 10, 2024).

16   The so-called Nigerian email scam, also known as the advance-fee scam, or the 419 scam (after the section of the Nigerian penal code dealing with fraud), is a type of scam in which the scammer promises the victim large sums of money, often received through inheritance following a tragedy much like the Barneses', but requires an advance fee or processing fee to initiate the transaction. Variations on the scam date to the late 18th century, but the modern version exploited the global spread of email technology. Though such scammers exist all around the world, this became known as a "Nigerian scam" because many early examples originated in Nigeria. A real-world example of such a scam email exploiting the Barnes tragedy and Flight 990 can be viewed, at least as of this writing, at <https://www.419scam.org/emails/2015-06/16/00875844.319.htm>. In that example, the scammer uses the real identities of William and Tamsin Barnes and the real details of their crash combined with a fictional story that they had no next of kin to receive their inheritance to bait the would-be victims.

17   Dunafon Castle (n.d.). About Us: A Rich History of the Castle. *Dunafon Castle*. <https://dunafon-castle.com/about-us> (accessed July 10, 2024).

18   *Ibid.*

tigation.

The caretaker claimed to have seen what he described as "energy" in the building and that the spirits had helped with the Castle's construction. Helpful spirits like those are always welcome as far as we're concerned, though we don't fully understand his meaning. Particularly with regard to seeing "energy," we have questions. Physicists define energy as the capacity to do work, which is not a visible thing (though the work itself might be). The word has been adopted by a lot of different types of spiritual movements, though, and in the process has lost much of its meaning outside of the scientific context. Sometimes it means something akin to what physicists talk about, but other times it might mean something more like "emotional state." In the context of the caretaker's story, we think he's probably talking about having seen some kind of entity in a non-humanoid form. Perhaps something similar to what's often described in paranormal photography as an orb or vortex (see Volume 1, Chapter 24).

Other stories included seeing a man in the Castle who wasn't really there, hearing footsteps throughout the Castle and even the sounds of large groups of people partying in the basement. Occasionally people outside the Castle could see shadows in the windows of people moving between rooms even when the Castle was vacant.

One of the Barnes sisters said that when she was growing up in the Castle, she would "practice playing dead" while lying in bed because she thought the pretense might fool whatever unseen entity haunted her room into leaving her alone. Both sisters said they could sometimes "feel" someone in the room with them and that it would even follow them into hallways even though no one was there.

Our Investigation

Several times, we were invited by Peter Boyles to conduct an investigation for the Halloween special of his KHOW radio program. In fact, if you've been reading this book in order you've already encountered some of those radio investigations. This was another one. It was conducted a few days before Halloween and consisted of a live radio broadcast followed by an investigation.

We arrived at the Castle around 6:00 p.m. and toured the place as usual before setting up for the radio show. The broadcast itself lasted from 7:00 to 9:00 p.m. and consisted of interviews with past and current Castle residents during which they talked both about the history of the property as well as their various ghost stories.

They even added to the ghostly lore live on air, one claiming that earlier that very day she'd seen the ghostly figure of a little boy on top of the Castle. Another claimed she'd been grabbed while walking around the property. Though we were present that day, we didn't witness these events for ourselves, so we can't really say too much about them.

When the broadcast ended, most of the people involved left for the night, a few retired to their bedrooms on-site, and we started setting up our own equipment for a more traditional investigation.

Following our usual tradition (why mess with what works?), we set up monitoring equipment throughout the Castle. In the hallway from the garage, we placed a video camera, microphone, remote thermometer, and a digital audio recorder. For the main living room, we set up two video cameras and a microphone. The hallway between the bedroom and the bathroom and the wine cellar each got one video camera and one microphone. Two of our team members also established posts, one in each location, to personally monitor the hallway from the garage and the living room.

Figure 12.3. The wine cellar. Photo: Bryan Bonner.

The investigation lasted for approximately six hours following our usual protocol. Most people had left or gone to bed for the night so we were able to work mostly as we usually do. But there were still other people around and we weren't always able to account for their movements, so it was sometimes difficult to draw conclusions. For example, we did hear a few unusual sounds—several slamming doors and a persistent "tapping" sound[19]—throughout the investigation that it would be tempting to chalk up to paranormal activity, but we simply don't know whether we heard ghosts or merely someone shuffling down the hall to use the bathroom.

EMF readings taken throughout the evening were at a level we expected from the property and did not vary unusually from one reading to the next.

All in all, it was a fun but pretty run of the mill kind of investigation.

But when we reconvened with the other guests the following morning, we got one more ghost story to add to the catalog. One of the people who'd stayed at the Castle said his girlfriend woke up in the middle of the night, said "they are all around us," and promptly went back to sleep as if nothing had happened.

---

19   This wasn't heard by the investigators on-site but we heard it in our recordings when we reviewed the data after the fact.

A ghostly visitation? Merely talking in her sleep? We have no way of knowing. For that matter, we don't have any way to verify the incident even took place (though for what it's worth, we believed him when he told the story). If only we had monitoring and measuring equipment in the room at the time so we might have been able to see what actually happened! Alas, sometimes our cameras just happen to be in the wrong place at the wrong time.

Our ultimate conclusions were that we didn't detect anything spectacularly out of the ordinary. A few anomalous sounds—and that story from another guest—caught our attention but lack of control during the investigation meant we weren't able to offer any definitive conclusions. But we'd certainly like to go back to follow up, both to see if we can catch any ghosts and simply because it's such a remarkable Castle.

References & Further Reading

ASPRS (n.d.). In Memoriam Archive: Tamsin G. Barnes 1934-1999. *American Society for Photogrammetry and Remote Sensing*. <https://www.asprs.org/a/news/archive/memoriam.html#Barnes>

Borger, J. & Dawoud, K. (2000). Wings and a prayer: EgyptAir flight 990 crash: special report. *The Guardian*. <https://www.theguardian.com/world/2000/may/08/egyptaircrash.usa>

Denver Post (1999). Obituary: Tamsin Barnes, William Barnes, Paula Barnes: Flight 990 victims. *The Denver Post*. <https://extras.denverpost.com/news/obits/barn1118.htm>

Dunafon Castle (n.d.). About Us: A Rich History of the Castle. *Dunafon Castle*. <https://dunafon-castle.com/about-us>

ECAA (2001). Report of Investigation of Accident: EgyptAir Flight 990, October 31, 1999, Boeing 767-300ER SU-GAP, Atlantic Ocean-60 Miles Southeast of Nantucket Island. *Egyptian Civil Aviation Authority*. Published courtesy of NTSB and archived from the original at <https://web.archive.org/web/20110622102818/http://www.ntsb.gov/events/ea990/docket/ecaa_report.pdf>

Jefferson County Open Space (n.d.). Lair o' the Bear Park. *Jefferson County Open Space*. <https://www.jeffco.us/1254/Lair-o-the-Bear-Park>

Lahoud, N. (2022). *The Bin Laden Papers: How the Abbottabad Raid Revealed the Truth About al-Quaea, Its Leader, and His Family*. New Haven, CT: Yale University Press.

Miller, D. (2013). Lair O' the Bear and Dunafon Castle. *Impression Evergreen*. <https://www.impressionevergreen.com/2013/02/lair-o-bear-and-dunafon-castle.html>

NTSB (2002). Aircraft Accident Brief: EgyptAir Flight 990, Boeing 767-366ER, SU-GAP, 60 Miles South of Nantucket, Massachusetts, October 31, 1999. *National Transportation Safety Board*. <https://www.ntsb.gov/investigations/AccidentReports/Reports/AAB0201.pdf>

13

Colorado's Most Haunted Mansion: The Croke-Patterson Mansion

People often ask, during the question and answer portions of our public lectures, what the most haunted place in Colorado is. We always say there are three answers. The most *famous* haunt is easily the Stanley Hotel (Volume 1, Chapter 14); the place we've seen the most ourselves will be the subject of the next chapter. But the one with, as far as we can tell, the most "naturally occurring" ghost stories reported in the literature must be the Croke-Patterson Mansion (also known as the Croke-Patterson-Campbell Mansion or the Patterson Inn) located at 420 E. 11th Avenue in Denver's Capitol Hill neighborhood.

Currently functioning as a luxury inn, the 12,500 square foot mansion sits on a quarter-acre lot and boasts nine bedrooms, nine bathrooms, a grand ballroom, a living room, a library, a dining room, and—according to the tales—more ghost stories than we have space to report. And the stories get weird. We're talking "gateway to Purgatory" and "fridge full of cats" weird.

Figure 13.1. The Croke-Patterson Mansion. Photo: Bryan Bonner.

The History

The Croke-Patterson Mansion, sometimes also called the Croke-Patterson-Campbell Mansion and now operating as the luxurious Patterson Inn, has a long and storied history. In fact, though we pride ourselves on writing detailed historic treatments of all of these properties, it's passed through so many owners over the years we can't possibly report everything, so we'll have to just hit the highlights.

The "Croke" part of the mansion's name originated with Thomas B. Croke (1852-1939). Born to a farming family[1] on March 4, 1852 in Rock County, Wisconsin, Mr. Croke came to Colorado in 1874 to become a schoolteacher. But that wasn't to be. Instead, he accepted a position as a clerk for the Daniels and Fisher department store where hard work eventually earned him promotion to manager of the carpet department. But if you think it remarkable for someone in such a profession (noble though it may have been) to have enough money to have a mansion named for him, you're not wrong. He eventually left Daniels and Fisher to found his own store, Thomas B. Croke & Company, offering carpets, window shades, and other furnishings.[2]

Business, apparently, was good, and Mr. Croke knew how to use his money, using his profits to invest in real estate throughout Colorado and eventually becoming president of the Denver Reservoir and Irrigation Company and getting himself elected to the Colorado State Senate in the early 1910s. Through these various ventures, he amassed enough wealth to commission the construction of what would become the Croke-Patterson Mansion. It was designed by architect Isaac Hodgson based on the Chateau d'Azay le Rideau in the Loire River Valley in France and built between 1890 and 1891 for the princely sum of $18,000 (equivalent to a little over $600,000 in 2024, further illustrating that real estate inflation—especially in Denver—has far outpaced general inflation).

Sources vary a bit on what happened next. Legend says that Mr. Croke stepped into his new luxury mansion—arguably *the* mansion in Denver—and immediately turned around and walked out, never to return. That's the legend, anyway, and a good illustration of the paranormal claims already beginning to creep into our historical section—after all, what *did* spook Mr. Croke so much?—but what do the local historians have to say about the matter? According to Phil Goodstein, the legend might not be too far from reality, as Mr. Croke never lived in the house.[3] No one has any idea what he saw, heard, felt, or experienced to scare him so much. Or maybe he just had a change of heart and decided he didn't like the house. Seems extraordinary but we suppose stranger things have happened. On the other hand, historian Amy Zimmer tells an entirely different story.

---

1   Their original family name was spelled "Croak" but it was changed in the mid-1800s. Such alterations of spelling were common among immigrants at the time (the Croaks/Crokes were Irish immigrants), much to the dismay of genealogists trying to trace family records across the centuries.

2   Goodstein, P. (1996). *The Ghosts of Denver: Capitol Hill.* Denver, CO: New Social Publications.

3   *Ibid.*

According to her, Croke did move into the house and lived there with his two children for about six months until he moved out, likely while still grieving the death of his wife in 1887 followed by the death of his mother shortly thereafter.[4] Others have suggested his quick move-out was motivated by a major financial loss on the silver market causing him to realize he couldn't afford to keep it.[5]

Regardless of the exact duration of his stay and regardless of the reason for his departure, what is absolutely clear is that he didn't keep the house for long. Paranormal enthusiasts will of course claim a ghostly reason for Croke's move, but as far as we know he never publicly gave an explanation so that's pure speculation. However, Croke would continue to live a successful life in other lodgings until his death in 1939.

The way Croke got rid of the mansion is interesting in and of itself. Instead of a simple sale, he traded the property in 1893 to Thomas McDonald Patterson (1839-1916) in exchange for 1,440 acres of Patterson's ranchland adjacent to property Croke already owned.[6] The exchange included all the Mansion's furnishings. Patterson announced the transaction to his wife in a letter, apparently having made the exchange on a whim without consulting his other half.[7]

Thomas Patterson himself immigrated from Ireland to the United States with his family in 1849, served briefly in the Army during the Civil War, studied as a lawyer and was admitted to the bar in 1867, and began a political career with the Democratic Party in Denver in 1972. It was a successful career. He began as a district attorney but managed to get himself selected as a delegate to the Congress from the Colorado Territory in 1874, during which time he played a role in establishing Colorado's statehood in 1876. He then served in the House of Representatives from 1877 to 1879, then acquired the *Rocky Mountain News* in 1890, which he used as a mouthpiece for his Democratic views and counterpart to the rival *Denver Republican*.[8] Though a lifelong Democrat, Patterson was part of the movement to attract Populists to the Democratic Party, allowing him to become acquainted with the famed lawyer and Populist/Democratic orator William Jennings Bryan who, according to rumor, was a frequent visitor at the Croke-Patterson Mansion. But despite his hesitations about the corporate side of

---

4   Zimmer, A. (2016). Denver's Most Haunted House. *Colorado Virtual Library*. <https://www.coloradovirtuallibrary.org/resource-sharing/state-pubs-blog/denvers-most-haunted-house/> (accessed July 10, 2024).

Zimmer, A (n.d.). Croke-Patterson-Campbell Mansion. *Colorado Encyclopedia*. <https://coloradoencyclopedia.org/article/croke-patterson-campbell-mansion> (accessed July 10, 2024).

5   Leggett, A. A. & Leggett, J. A. (2011). *Haunted America: A Haunted History of Denver's Croke-Patterson Mansion*. Charleston, SC: The History Press.

6   Zimmer, A (n.d.). Croke-Patterson-Campbell Mansion. *Colorado Encyclopedia*. <https://coloradoencyclopedia.org/article/croke-patterson-campbell-mansion> (accessed July 10, 2024).

7   Leggett, A. A. & Leggett, J. A. (2011). *Haunted America: A Haunted History of Denver's Croke-Patterson Mansion*. Charleston, SC: The History Press.

8   Goodstein, P. (1996). *The Ghosts of Denver: Capitol Hill*. Denver, CO: New Social Publications.

the Democratic Party, Patterson re-entered national politics and served as United States Senator from Colorado from 1901 to 1907.[9] He eventually sold the *Rocky Mountain News* to Paul Schafer (who also bought the *Denver Republican* in the same year), but not before the rivalry between the *News* and *The Denver Post* came to a head in December of 1907 when Patterson accused *Post* owner Frederick G. Bonfils of blackmail in retaliation for which Bonfils attacked Patterson on the street.[10]

Patterson is also memorialized in United States Supreme Court history. He had written articles accusing the Colorado Supreme Court of upending election results for which he received a contempt citation which was upheld by the same Court. He appealed to the United States Supreme Court making a First Amendment argument as well as arguing that the state Supreme Court's ruling was self-serving. Writing for a 7-2 majority, famed Justice Oliver Wendell Holmes, Jr. upheld the contempt citation arguing that the First Amendment did not apply to the states and so the questions involved in the case were a matter of local law, but that even if the First Amendment did apply, its purpose was only to prevent prior restraint.[11] Justices John Marshall Harlan and David J. Brewer dissented, the former arguing that the First Amendment did apply to the states through the Fourteenth Amendment and the latter on jurisdictional grounds. Though Harlan's argument didn't win the day in 1907, the Court did eventually adopt his argument, now commonly known as the "incorporation doctrine," in 1925's Gitlow v. New York which held that the due process clause *does* apply the First Amendment to the states.[12]

Senator Patterson's life was marked by extremes of both success and tragedy. He was clearly wealthy and successful in both business and politics and by all accounts a devoted (if not always present) husband to his wife, the noted suffragist Katherine Grafton Patterson (1839-1902) and his family. But he also suffered numerous humiliating setbacks and personal tragedies. His son committed suicide at the age of twenty-six years in 1892. Just two years later, he lost his daughter to an unspecified illness at the age of twenty-seven years. And Katherine Grafton Patterson died in their home on July 16, 1902 "from nervous prostration following a violent attack of acute cholera morbus."[13]

Returning to our main story, the Pattersons made good use of the Mansion for their growing family. Richard C. Campbell (1869-1930), another newspaperman, moved to Colorado in 1894 where he met Patterson's daughter Margaret. Soon the two were wed and lived in the Mansion with the Patterson family along

---

9   *Ibid.*

10   Denver Post (2009). Rocky Mountain News history timeline. *The Denver Post.* <https://www.denverpost.com/2009/02/26/rocky-mountain-news-history-timeline> (accessed July 10, 2024).

11   Patterson v. Colorado, 205 U.S. 454 (1907).

12   The rest of Holmes' reasoning was superseded in 1931's Near v. Minnesota which held that the First Amendment's protections are not limited to the mere prevention of prior restraint.

13   Indianapolis Journal (July 17, 1902). Obituary. Archived at <http://ingenweb.org/inmontgomery/obits%20pa/patterson,-katherine.html> (accessed July 10, 2024).

with their own three children.[14] Campbell was a lifelong Republican and so often found himself at odds with his father-in-law's politics, but their personal relationship was strong enough that Patterson willed his Mansion to Campbell and Margaret Patterson-Campbell upon his death in 1916.[15] This is why some think the Croke-Patterson Mansion should more properly be called the Croke-Patterson-Campbell Mansion.[16]

The Campbells lived in the Mansion for a time until they moved to a more modern home on York Street in 1924, where they lived until Margaret died in June of 1929 followed shortly by Richard's death in February of 1930.[17] Through the years following the sale, the Mansion passed through numerous hands. Among other purposes, it was home to the Joe Mann School of Orchestra, the KFVR radio station, and an apartment complex. Other than the Pattersons, no one seemed to stay very long.

The next noteworthy owners were Dr. Archer Sudan (1892-1971) and Tulleen (often misspelled "Tuleen"[18]) Swift Sudan, who purchased the Mansion in 1947 and held it, albeit not happily, until 1958.[19] Dr. Sudan was an accomplished physician who began his career as an intern at Denver General Hospital, where he met and fell in love with a nurse named Tulleen Swift, whom he later married. His career was a successful one and he served as president of the Colorado Medical Society from 1946-1947.[20] Following that service, Dr. Sudan, as such a prominent and respected physician, was often away for extended periods giving medical training and other professional appearances, leaving Tulleen alone in their huge home (their son, Archer, Jr., did not live with them at that time).

Nevertheless, Tulleen was supposed to be a happy person who enjoyed her own social circles until a miscarriage led her into a deep depression. According to legend and rumor (all quite unsubstantiated but also quite relevant to the forthcoming ghost stories), she was so distraught by the tragedy that she buried the baby in the walls of the Mansion before taking her own life in 1950 by placing

14   Zimmer, A (n.d.). Croke-Patterson-Campbell Mansion. *Colorado Encyclopedia.* <https://coloradoencyclopedia.org/article/croke-patterson-campbell-mansion> (accessed July 10, 2024).

15   Goodstein, P. (1996). *The Ghosts of Denver: Capitol Hill.* Denver, CO: New Social Publications.

16   We choose to use the shorter version honoring only the home's builder and first long-term residents, because if we start listing all of the owners of the property, the Mansion's name would end up being dozens of words long, as you'll see as we continue to trace its history.

17   Zimmer, A (n.d.). Croke-Patterson-Campbell Mansion. *Colorado Encyclopedia.* <https://coloradoencyclopedia.org/article/croke-patterson-campbell-mansion> (accessed July 10, 2024).

18   Records are sparse, so we've assumed the spelling on her grave marker (with two Ls) is the correct one.

19   Goodstein, P. (1996). *The Ghosts of Denver: Capitol Hill.* Denver, CO: New Social Publications.

20   Wier, J. & Miller, J. (1997). Of Things Medical in Middle Park. *Grand County Historical Association Journal, 14*(1).

cyanogas (a powdered form of rat poison also known as calcium cyanide) into a hot bathtub to create deadly cyanide gas.[21] Further exploration of this episode will be included in the "Our Investigation" section shortly.

History hasn't recorded many details about how Dr. Sudan coped with the loss, but it is known that five years after Tulleen's death, he was remarried to his longtime friend Martha Hawkins with whom he remained until his death in 1971.[22] However, he didn't spend his final years at the Croke-Patterson Mansion. In either 1957 or 1958, the Sudans moved out of the house and spent the rest of their years in Lafayette. The property remained within the family, though, and Archer Sudan, Jr. moved in serving as a live-in landlord for apartments established in the Mansion's rooms until he sold it in 1972.[23] It was around this time that the paranormal claims started to crop up, but we'll get there shortly.

During the early 1970s, a quiet battle raged between developers and historic preservationists. Perhaps that battle is always raging, but in this case it focused on Denver's Capitol Hill neighborhood. The 1972 destruction of the nearby Moffat Mansion led to a grassroots effort to preserve as many of Denver's historic properties as possible. The crumbling Croke-Patterson Mansion was surely going to be a target for destruction because it sits on prime real estate, but local realtor Mary Rae fell in love with the property and purchased it from Sudan in 1972, saving it from destruction and planning to continue operating it as an apartment building.[24] Purchasing the property alone might not have been enough to ensure its salvation. Maintenance and utilities for such a large and old property are substantial expenses, and there's always the risk an owner may need to sell to someone else—and in any real estate transaction, there's little to guarantee subsequent owners will have the same respect for historic preservation. But in 1973, Mary Rae and her husband John succeeded in acquiring protection for the building by listing it both as a Denver Landmark[25] and on the National Register of Historic Places.[26]

Rae didn't stay long. She and John sold the place in 1976, at which point it went through a variety of hands, existing variously as either apartments or office

21   Lamb, K. (2016). *Ghosthunting Colorado*. Covington, KY: Clerisy Press.

22   Wier, J. & Miller, J. (1997). Of Things Medical in Middle Park. *Grand County Historical Association Journal, 14*(1).

23   Lamb, K. (2016). *Ghosthunting Colorado*. Covington, KY: Clerisy Press. Goodstein, P. (1996). *The Ghosts of Denver: Capitol Hill*. Denver, CO: New Social Publications.

24   Zimmer, A (n.d.). Croke-Patterson-Campbell Mansion. *Colorado Encyclopedia*. <https://coloradoencyclopedia.org/article/croke-patterson-campbell-mansion> (accessed July 10, 2024).

25   City and County of Denver (2023). Denver Individual Landmarks. City and County of Denver. <https://www.denvergov.org/files/assets/public/v/4/community-planning-and-development/documents/landmark-preservation/individual_landmarks_list.pdf> (accessed June 18, 2024).

26   Emrick, C. (1973). Croke-Patterson-Campbell Mansion: National Register of Historic Places Inventory – Nomination Form. [Register No. 73000467]. *National Register of Historic Places*. <https://catalog.archives.gov/id/84129551> (accessed July 11, 2024).

space, and during which time the ghost stories and unusual occurrences (which we're getting to soon, we promise) began to stack up.

In or around 1998, the property was again purchased by Dr. Douglas E. Ikeler, a veterinarian, and his wife Melodee who rebranded the building "Ikeler Castle" and remained for about 10 years. It wasn't, apparently, the happiest of decades for the Ikelers. Though things seemed to be going well from the outside—Dr. Ikeler had a successful veterinary practice and the couple started their family with triplets while living at the house—they were plagued by alleged paranormal happenings. Scandal isn't our business so we'll not repeat all of the details here, but their marriage ended in an acrimonious divorce including contentious custody battles which made it all the way to the Colorado Supreme Court[27] and their ownership of the Croke-Patterson Mansion[28] ended in foreclosure in 2007. Though once quite wealthy, apparently Dr. Ikeler couldn't afford both the cost of the divorce and the cost of maintaining the property which he was claimed, cost some $100,000 for the mortgage and an equivalent amount for utilities and taxes.

Indeed, at one point some of our members almost considered looking into buying the Mansion when it went for sale some years back. The asking price, particularly by Denver standards and for such a large property, was almost absurdly low. But we learned it's not the price of the property itself that gets you. It's all the other costs involved in keeping it habitable.

Dr. Ikeler is still alive and now writes books instructing people about how they can live as long as 150 years.[29]

Figure 13.2. A crumbling (now repaired) façade. Photo: Bryan Bonner.

---

27   In re Marriage of Ikeler, 161 P.3d 663, 667 (Colo. 2007) (en banc).
28   We refuse to call it "Ikeler Castle." For one thing, we prefer to honor its historic owners. For another, it's not a castle; it's a mansion. "Castle" has a specific definition which this property doesn't meet.
29   Ikeler, D. (2008). *Life Extension: How to Use Your Neurophysiology to Live 150 Years!* [self-published].

The Mansion then remained in the custody of financial institutions until it was purchased for only $565,000[30] in 2011 by filmmaker and architect Brian Higgins who planned to make a documentary film detailing the restoration process. The film was released in 2013, the same year the property opened for business in its current incarnation as Patterson Inn.[31] As for the documentary, it ended up being a lot more paranormal than your average architectural film, but we'll get to that. The Patterson Inn now offers adults-only accommodations in nine themed bedrooms in the fully-renovated building including a complimentary breakfast and common areas (including a delightful reading room) that are a history buff's dream.[32]

Why has the Croke-Patterson Mansion passed through so many owners over the years? Could it simply be the cost of keeping up such a large property? Or could something else be going on? It's to that question we now (finally) turn our attention.

Paranormal Claims

It's difficult to know where to even begin reporting all the paranormal claims that have been made of the Croke-Patterson Mansion. Try as we might, there's simply no way we can cover them all. Not only is the building more than a century old (more than enough time to accumulate its share of folklore), but it's been home to some remarkable characters throughout history, its style simply *looks* haunted, and it's by this point been featured in numerous books, websites, television programs, documentaries, and more, all describing its haunted reputation. A good place to begin, we suppose, would be at the beginning.

The very beginning, we've already touched upon while discussing the history. Historians may disagree regarding whether Mr. Croke set foot in his new home for the first time and immediately turned tail never to return or whether he lived there for some six months, but it's well-established that he didn't stay long. At the time, that probably didn't seem like anything more than odd or eccentric behavior, but in the decades since, many have mused that perhaps he was scared off by something supernatural. Maybe he saw a ghost. Perhaps he felt something evil. Whether an entity was in the house or something was "wrong" with the house itself—both claims will be made again as we continue to explore the stories—the legend goes that Croke was wise enough to get out while he could.

And that legend has only grown since then. Many have claimed that, whatever supernatural thing or things might be happening within those walls, it or they are not good for the people who live there. Could the house and/or its ghosts be to blame for the various illnesses, deaths, marital disputes, and other misfor-

---

30   We told you it was cheap. That may be a lot of money, but it's not "gigantic historic mansion in the heart of Denver" money. The low price was due to disrepair.

31   Zimmer, A (n.d.). Croke-Patterson-Campbell Mansion. *Colorado Encyclopedia.* <https://coloradoencyclopedia.org/article/croke-patterson-campbell-mansion> (accessed July 10, 2024).

32   Patterson Inn (n.d.). About the Inn. *Patterson Inn.* <https://www.pattersoninn.com/about/> (accessed July 11, 2024).

tunes that have plagued its residents? One former resident even cautioned a new resident not to stay there, adding that he hoped the new resident wasn't happily married if he was going to spend any time in the house.[33]

According to Phil Goodstein, though, it wasn't until after Dr. Sudan moved out that the Mansion began to develop its paranormal reputation in earnest. Building on the story of Tulleen Sudan's miscarriage and suicide, legends began to crop up that cribs would swing on their own, objects would fly through the house, and more.[34] Presumably these occurrences were the work of either Tulleen herself or the ghost of her ill-fated child.

Melodee Ikeler shared a firsthand story along these lines with us. Toward the end of her pregnancy once day she was lying in bed and struggling to get up. Recall she was about to have triplets, so the struggle is more than understandable. While she was trying to roll out of bed, she looked up and saw a woman standing next to the bed offering a hand of assistance. She took the woman's hand and pulled herself out of the bed, then the woman simply vanished through the wall. She didn't identify the woman specifically as the ghost of Tulleen Sudan—she likely wouldn't have known what Tulleen looked like anyway—but subsequent lore has connected the story to the kind intervention of the ghostly mother who lost her own child.

Details of Tulleen's own story vary a bit. It's known she died, probably of suicide, in 1950, but different versions of the ghost story vary in the details. Most commonly, she's said to have killed herself after miscarrying and so her ghost is thought to haunt the Mansion. Fair enough. But others have added on that prior to her suicide, she actually buried the miscarried baby in the walls of her home. Another rumor suggests a child was actually tortured to death somewhere in the house, either in the basement or the carriage house. At one point, a psychic even told investigators where to dig in the walls to find the child's body, though we don't spoil much of our own investigation to point out that they never found anything after knocking a hole in the wall.[35]

Ghost stories might not have become popular until after the Sudan residency, but ghostly activity has also been attributed to prior residents. A commonly reported occurrence in the Mansion's history was that typewriters would suddenly make a racket when no one was present, as if some unseen force or entity were trying to write a message. Our favorite interpretation of this phenomenon ascribes it to the ghost of Senator Patterson. Recall that in addition to being a politician, he was a newspaper owner. This legend suggests that whenever a political scandal goes unreported in the press, his ghost becomes so agitated that he turns on all the communications equipment—typewriters, printers, phones, computers, you name it—in his attempt to communicate the news to the living world.[36] We can only imagine what he must think of some of the political scandals these days.

---

33   Higgins, B., Marcus, C., & Salvione, E. (Producers), & Higgins, B. (Directors). (2013). *The Castle Project* [DVD].

34   Goodstein, P. (1996). *The Ghosts of Denver: Capitol Hill.* Denver, CO: New Social Publications.

35   *Ibid.*

36   *Ibid.*

These tales were particularly common during the time the Mansion operated as an office space.[37] Additional reports claim that Patterson's ghost has sometimes been seen to visibly manifest, particularly in the courtyard.

Figure 13.3. No infant remains were found within these walls. Photo: Bryan Bonner.

We've also heard reports of a ghost named "Maggie" who is said to visit the house on particularly cold nights. We can't be sure whose ghost this is meant to be, but Maggie is the diminutive form of Margaret, so it could be Margaret Patterson Campbell. We're not sure if she ever went by that nickname or not.

Other tales are not so specifically connected to the Mansion's known history or owners, but that doesn't mean we're anywhere close to finished with our catalog of the ghost stories. Many of them are simply the stereotypical reports you hear from any famous haunt. Lights flickers, objects move, doors open and close, witnesses see shadow people either wandering the halls or standing in the windows. With regard to those stories, the only thing that sets the Croke-Patterson Mansion apart from all the other haunts is the sheer volume of lore associated with it. Because it's been not only a private residence but also apartments, office space, and a hotel, it doesn't take long as all to find someone in the Denver area with a ghostly story of their own, and many of them are more than willing to share. Those who haven't personally experienced anything paranormal are never short of published accounts to read.

One of the unidentified ghosts we've heard about is an "old man" who sometimes wanders between the Mansion and the carriage house. Others include

37   Lamb, K. (2016). *Ghosthunting Colorado*. Covington, KY: Clerisy Press.

a woman who hanged herself in the turret room, a man who hanged himself in the basement, and a whole host of unknown characters.

One that received a bit more attention was a spirit named "Willie" (sometimes spelled "Willy"). According to the rumor, he was one of the men ordered to dig up the graves at what was to become Cheesman Park (see Volume 1, Chapter 5 for that whole sordid affair). But instead of just relocating the bodies, he was robbing the graves.[38] Worse, he was taking the bodies themselves for use in Satanic rituals. Still worse, he eventually got bored of using the bodies for his dark magick and started kidnapping people off the street. He even killed a young boy and hung him from a tree in front of the carriage house in one version of the story, though more common versions have him killing multiple people by hanging them from a steel beam that sticks out of a wall in the carriage house, then secretly moving the bodies back to the Mansion proper by means of a tunnel (which has since been sealed) between the two buildings.

During Mary Rae's ownership of the property, she received numerous complaints from her tenants about a particular room on the fourth floor of the Mansion. Multiple people reported hearing sounds ranging from a screaming baby to a loud party coming from that one room. However, not only was the room not occupied by any other tenants, it was actually just a small storage room. Mary never figured out what the sounds were. An anonymous source who claimed his/her mother lived at the Mansion in the 1970s reported a similar claim of loud parties, but located it on the third floor instead of the fourth.

And then there are the suicidal dogs. Bear with us while we lay out the background. Throughout the Mansion's history, break-ins and burglaries have been a problem. Large houses, unoccupied and falling into disrepair as the Croke-Patterson Mansion was several times in its history, are unfortunately ripe targets for criminals. The largest of these burglaries happened in 1976. Neighbors noticed the Mansion's back door had been forced open and police responded to the call. They found six of the large stained glass windows had been removed from their settings and leaned against the back wall. When police left after failing to find any suspects, the criminals returned the next day to complete the theft.

Other crimes happened during various periods of renovation. For example, when the building was being converted to office spaces, construction workers often showed up in the morning to find tools missing or even that the work they'd done the previous day have been undone overnight. Sometimes this is attributed to ghostly activity and sometimes to criminal activity. Use your own judgement since we weren't present to investigate, but we have to admit we suspect a combination of criminal theft to explain the missing tools and lazy workers to explain the work that turned out not to be done the next morning.

---

38   In case you haven't read our first volume yet, the very abridged version is that men were hired to relocate the graves, and people have accused these workers—particularly undertaker McGovern—of all kinds of abuse including grave robbing. The truth is that McGovern was just an innocent man hired to do a dirty job, but because a lot of graves were being dug up, some grave robbery did occur. It's unclear whether any of McGovern's hired men participated in the crimes. It's also unclear whether any of McGovern's assistants were named Willie.

Homeless people also sometimes took shelter in the Mansion, again most often during periods of renovation.

With all of this going on, the owners of the building needed reliable security. Unfortunately, at least according to the legend, they seemed to be unable to find anyone. Guards would abandon their posts in the middle of the night without notice, never to return. One of them only lasted a single day. Something was scaring them off. The owner tried installing a security fence but it wasn't up to the job of protecting the property and homeless people and thieves still found their way in.

Enter the dogs. Figuring they'd be more loyal or reliable than hired guards or fences, the owner acquired guard dogs for the property. And here's where reports vary a little bit. Some reports say there were two German Shepherds.[39] Others claim there were three Dobermans.[40] Our own mental image alternated between the two so often we ended up settling on Doberman Shepherd mixes (or "Shedobies.") What they all agree on is that something terrifying happened with the dogs. The first one, all the reports agree, leapt through the third-floor turret window, killing itself instantly. No one knows what scared it so much that it would autodefenestrate, though subsequent visitors have reported an uneasy feeling in that turret room.

Figure 13.4. The turret window from which a guard dog was allegedly defenestrated. Photo: Bryan Bonner.

What happened next is also a bit unclear. In one version of the story, the second dog was found catatonic the next night. In the other version, the second

39   Higgins, B., Marcus, C., & Salvione, E. (Producers), & Higgins, B. (Directors). (2013). *The Castle Project* [DVD].

40   Goodstein, P. (1996). *The Ghosts of Denver: Capitol Hill*. Denver, CO: New Social Publications.

dog killed itself in a similar manner to the first one, and only *then* was the third found catatonic.

A small but intriguing paranormal claim surrounding the property is that mysterious puddles of water appear on the sidewalk just beyond the front door even when it hasn't rained or snowed in a long time. Various spiritual claims are made of water. Religious people have baptisms and holy water. Vampires are supposed to be unable to cross running water. Ghosts are alternately said to be either attracted to or repelled by water depending on whose lore you're reading. Reports don't seem to agree on an interpretation as to what the Croke-Patterson water might mean, but it is often seen on the grounds.

Perhaps the most grandiose of the paranormal claims surrounding the Mansion was proposed in the 2013 documentary *The Castle Project.* Full disclosure: some of our current and former members were interviewed in the documentary, but we didn't expect the claims to go in the direction they ultimately did. The film starts off by simply documenting the renovation process, discussing the history, and exploring the ghost stories just like any other documentary might. But by the end, it proposes (hypothetically but strongly) that the Mansion is so full of spirits because it's a gateway to Purgatory.[41]

Purgatory is a sort of level of the afterlife in Catholic doctrine.[42] Essentially, it's where people go after death if they're not damned to Hell but not yet pure enough for Heaven. It's a place of purifying suffering, though not thought to be permanent; souls in Purgatory will eventually reach Heaven once they've been sufficiently purified.

The documentary observes that artistic representations of Purgatory—particularly Gustave Doré's illustrations of Dante's *Purgatorio*—bear a striking resemblance to the rock structures of the Garden of the Gods and the Manitou Red Rock quarries from which the stone for the Mansion's construction was obtained. Combined with the fact that "Manitou" means "spirit," this led the documentarians to posit that the stone itself is spiritually significant and turned the Mansion into a kind of waystation for spirits or a portal into Purgatory. Therefore it's so haunted not because of the people who've lived and died in the Mansion but because of that religious significance.

During the documentary's production, a fire broke out in the Mansion, destroying much of the work that had been done. The documentary hints that this may have been due to some ghostly cause, claiming the fire stared at the very same moment as a wildfire miles away. Other sources have jumped on the paranormal interpretation of the fire as well.

For what it's worth, other witnesses have claimed to experience spirits not of the Mansion's prior residents but of people from more recent history, which

---

41   Higgins, B., Marcus, C., & Salvione, E. (Producers), & Higgins, B. (Directors). (2013). *The Castle Project* [DVD].

42   Some other religions have similar ideas. Orthodox Churches often follow Catholic teachings. Judaism has a concept of "Gehenna." Islam has various levels of Hell, the highest of which is similar to Purgatory. Hinduism has the idea of Naraka. All of these are similar in some respects to the Catholic Purgatory. Most protestant denominations reject the doctrine.

does at least somewhat seem to match up with the documentary's Purgatorial claims. But it's at this point we have to start jumping in with our own commentary, so we'll close this section and transition over to our own investigative notes.

Our Investigation

Our investigations into the paranormal phenomena at the Croke-Patterson Mansion span several years. In that time we've been present for investigations during radio programs (much like the ones in the preceding chapters). We've helped host public events, conducted private investigations on our own, spent time working on historic research, been interviewed for a variety of film, radio, and television programs on-site at the Mansion, and more. Because this investigation was so spread-out over the course of multiple visits, it doesn't make much sense to provide a linear recounting of the process. Instead, we'll try to tackle this according to topic.

Since it's where we closed our previous section, let's start here with the claim that the Mansion might be a kind of gateway or window into Purgatory. To us, that's the most extraordinary of all the claims made about the property (though this place offers an embarrassment of riches when it comes to extraordinary claims). It's also the one we're probably least able to forensically investigate because it's directly related to religious doctrine (and as we always remind people, we're happy to explain what different religions believe but in the absence of objective evidence, we remain neutral on religious philosophy).

What we can say is that the striking red sandstone from which the Mansion was constructed was indeed quarried from the Kenmuir Quarry in Red Rock Canyon.[43] Near Manitou Springs, Colorado, this location (now operated as city park) is an extension of the same geological Fountain and Lyons formations that make Garden of the Gods so spectacularly scenic. To avoid repeating ourselves too much, we'll refer you to Volume 1, Chapters 6 and 11 of this series for more detail on these geological formations, but we'll remind you that the Manitou Springs area is named for the concept of the "manitou," which in Algonquian traditions refers to a type of spirit or life force. However, simply observing, as the documentary did, that there's some spiritual significance to the stone only because the name "Manitou" is related to the word "spirit" is unconvincing to us, particularly in light of the fact that the manitou is an Algonquian concept and the various claims regarding Croke-Patterson Mansion and Purgatory are of Christian origin.[44]

Indeed, the documentary makes much of Christian symbolism found

---

43   Friends of Red Rock Canyon (n.d.). Red Rock Canyon Stone in Denver's Historic Buildings. *Red Rock Canyon Open Space.* <https://redrockcanyonopenspace.org/education/history/red-rock-canyon-stone-in-denvers-historic-buildings/> (accessed July 11, 2024).

44   We don't rule out the possibility that both religious traditions could be using different language to express the same underlying reality, but we're not going to hang our hats on that kind of speculation either.

throughout the building's architecture and décor.[45] A few of the examples are less convincing than some others, but we agree: there is a lot of religious symbolism built into the design. However, we also point out that a *lot* of architecture borrowed elements from religious iconography, both to honor religious faiths and simply because designers liked some of the symbols. The unfortunate trend in recent decades favoring brutalist architectural designs[46] notwithstanding, architectural history is full of artistic symbolism borrowed from a variety of religious sources (though primarily Judeo-Christian sources in the West). Especially since Croke-Patterson Mansion was designed specifically to take inspiration from another historic property, we'd be surprised if some Christian symbology, as part of the Western cultural tradition, *didn't* find its way in.

And pretty much the same can be said of the overarching claim that the red sandstone formations bear resemblance to images of Purgatory. Whatever your religious tradition may be, our ideas of what Purgatory might look like (if one believes in it) come only from artistic interpretations of religious doctrine, not from some divine revelation itself. French artist Gustave Doré was an undeniable genius[47] and his illustrations of Dante's *Divine Comedy* have become inextricably linked with the poem itself—and yes, some of his illustrations do bear resemblance to geological formations. But there's nothing to suggest that Doré was inspired by anything other than genius-level artistic vision, likely combined with whatever geological formations he'd seen in his own life.

In other words, we have no way of saying the Croke-Patterson Mansion *isn't* the gateway to the afterlife. But the claim is based on reasoning we simply find unconvincing.

With regard to the fire that happened during production of the *Castle Project* documentary, it really did happen. But rather than a paranormal cause, it was started when a pile of oily rags on the second-floor landing spontaneously combusted. One wouldn't necessarily expect this to happen absent some kind of spark or ignition source, but any woodworker can tell you it really does happen. The reason is that certain types of oils (such as linseed oil used in wood finishing) oxidize as they cure. That chemical process releases a small amount of heat. But if the rags are in a pile and the heat can't just disperse into the environment, the temperature of the rag pile gradually increases until it reaches combustion temperature.

Mysterious puddles of water seen on the property also have an explanation, and it's one that connects to several of the characters we've already met both in this chapter and in our previous volume. Locals don't need a detailed book on history, geography, or meteorology to know that Denver is a fairly arid region. Early in its history, it needed a reliable way to move water from the South Platte River to the various neighborhoods that needed the water because waiting for rainfall

---

45   Higgins, B., Marcus, C., & Salvione, E. (Producers), & Higgins, B. (Directors). (2013). *The Castle Project* [DVD].

46   Ick.

47   Colorado locals or visitors are strongly encouraged to view his "La Famille du Saltimbanque: L'Enfant Blessé" (The Family of Street Acrobats: The Injured Child) at the Denver Art Museum. It's haunting and magnificent.

simply wasn't going to be an option. The solution was to build systems of dams, reservoirs, and—most importantly—irrigation canals.

Heroes of Denver's water infrastructure included none other than Mr. Croke himself, who co-founded and served for a time as president of the Denver Reservoir and Irrigation Company.[48] Walter Scott Cheesman (see Volume 1, Chapter 5) was known as Denver's first "water baron." But the man most responsible for bringing water to the Capitol Hill neighborhood was John W. Smith. In 1864, he got the job of digging a 25 mile ditch to divert water from the South Platte to Capitol Hill.[49] Not only was he successful, but his creation, known various as the Smith Ditch, the Big Ditch, the City Ditch, or the Denver Ditch, still exists and is known as the "oldest working thing" in Denver.[50] At this point, most of the Ditch and its spiderweb of offshoots are underground, but a visible portion can still be visited in Washington Park.

Why are we suddenly talking about the history of Denver irrigation systems in the middle of our ghost stories? It turns out one of the offshoots of the Denver Ditch still passes under the Croke-Patterson Mansion.[51] From time to time—not every day by any means, but often enough to be noticeable—some water bubbles up to the surface from the subterranean irrigation system and wets the sidewalk. Over the years, people have attempted to patch up the cracks from which the water flows, but water always seems to win and it just moves a little further up the sidewalk to reemerge from a new spot.

Claims surrounding Tulleen Sudan are also difficult to validate. At no time during our investigations did we witness any phenomenon which might be attributed to her spirit, but that doesn't mean she doesn't sometimes appear to other people, so we have to leave that largely as an open question in our books. It is worth noting, though, that claims she buried her child in the building's walls seem to be false. No such remains have ever been found, even during extensive renovation projects.

Whether she actually had a miscarriage and committed suicide is a more difficult question to answer. Cataloguer of local ghost stories Kaitlyn Lamb points out both that no one has found any records of a birth or miscarriage in the Mansion and that Mrs. Sudan would have been forty-seven years old at the time of her death, past the age of healthy childbearing.[52] Though that does raise questions about the story's veracity, it doesn't rule it out entirely. Women have had children at more advanced ages than that, after all, and it's entirely possible, furthermore, that she could have miscarried due to her age. But the lack of any records to substantiate the claim should raise our skeptical antennae a bit.

What we do know is that she did actually commit suicide in 1950. Our ini-

48   Goodstein, P. (1996). *The Ghosts of Denver: Capitol Hill*. Denver, CO: New Social Publications.

49   *Ibid.*

50   Peterson, E. (2023). Only in Denver: City Ditch, the 'Oldest Working Thing.' *Visit Denver*. <https://www.denver.org/blog/post/city-ditch/> (accessed July 11, 2024).

51   Goodstein, P. (1996). *The Ghosts of Denver: Capitol Hill*. Denver, CO: New Social Publications.

52   Lamb, K. (2016). *Ghosthunting Colorado*. Covington, KY: Clerisy Press.

tial thought, given our inability to find public records or obituaries for Mrs. Sudan, was that the entire story might have been merely the stuff of urban legend. However, haunted history authors Ann Alexander Leggett and Jordan Alexander Leggett managed to accidentally unearth a copy of a death certificate for Tulleen Sudan misfiled in a box of real estate records indicating that she did indeed die of self-inflicted cyanogas poisoning.[53]

Assuming the story of the miscarriage is mere rumor (which seems to be the safest assumption to make), the cause of her depression is uncertain. It could very well be that despite her prior ability to happily socialize during Dr. Sudan's frequent absences, she eventually sunk into a depression because of his absence and her relative isolation. If so, her story has an even more tragic end. Likely because Dr. Sudan remarried after her death, her grave in Mount Olivet Cemetery remains a solitary one.[54]

So much for background research. Let's examine some of our findings from our various on-site investigations. These were conducted at various times and under various conditions, giving us a good opportunity to experience the house in different seasons and different times of day, both alone and with others present. That turned out to be fortunate because some of our discoveries are dependent on the conditions at the time.

One of the things we wanted to look into was the case of the suicidal dogs. We weren't present during the incident itself but it was so commonly repeated we wanted to see if we could learn anything about it. So on one of our investigations we made sure to wander up to the turret window out of which the first dog was supposed to have taken a death plunge. What we found made us even more skeptical.

The turret room is a tiny little room accessible by means of a staircase. We reasoned that in order to jump with enough force to break through the turret window's glass, even a full grown human being (and all the more true of a dog) would have to get a running start. But there simply isn't enough room to do so. Any dog who jumped into the window from a stationary position in the room (as it would have had to do because of the space) most likely would have just bounced off the glass and emerged with nothing worse than a headache.

Some people have suggested the origin of the story might have been that a homeless man trying to take shelter in the Mansion threw a dog out of a first-floor window or that a dog might have jumped (safely) from that window. This is possible, but we're not sure whether or not it happened. The number of reports sharing this story, some of which even include eyewitness statements, suggest this could be the kernel of truth behind the legend. Dog lovers can rest easy knowing that, whether they were German Shepherds or Dobermans, there's no reason to think anything bad happened to the pups.

Also curious about the "Willie the hangman" story, we spent some time checking out the carriage house. Historic research already left us pretty skeptical of the whole story. Without blowing our own horn too hard, we'd like to think

---

53   Leggett, A. A. & Leggett, J. A. (2011). *Haunted America: A Haunted History of Denver's Croke-Patterson Mansion.* Charleston, SC: The History Press.

54   *Ibid.*

we're among the world's experts on the history of Cheesman Park and the City Cemetery relocation and we'd never heard any stories resembling that of Willie. We were also unable to find any police records or newspaper articles to corroborate the tale. Absence of evidence sometimes really is evidence of absence (though perhaps not *proof* of absence), but we still wanted to check out the site.

Figure 13.5. The "hangman's beam" was part of a winch system. Photo: Bryan Bonner.

As it turns out, there really is a piece of metal sticking out from the wall, just as the urban legend suggests Willie used to hang his victims. Far from being a "hangman's beam," though, it was actually just part of a winch and pulley system used to carry heavy supplies to the second floor when it was an active carriage house.

Does the lack of historic corroboration and the explanation for the metal beam disprove the Willie story? It does not. It's possible someone could have been hanged from that winch and pulley system and it's possible for historic records to be incomplete. But the question you have to ask yourself is: if there's no record of the incident in the historic documents, where did the story come from in the first place? Mere urban legend is the simplest and best explanation.

Most of our on-site investigations, fun though they absolutely were for a variety of reasons, were less than interesting on the paranormal level. As is our way, we utilized continuous video and audio monitoring as well as periodic EMF and seismic measurements throughout the several investigations. Between all of our trips out there, we covered if not every portion of the Mansion, at least the vast majority of it. And we never found any anomalies on our various gizmos. But that doesn't mean we always came up empty, either.

One of the investigations took place in the dead of winter. At the time, the entire building had been "winterized." That means the heat and running water had been shut off, so we had to improvise and keep ourselves alive with some space heaters we brought from home. For some reason, it's become a tradition for us to conduct investigations in freezing cold temperatures, sometimes in unheated venues, often during a blizzard. It's not planned that way; that's just how things shake out sometimes. Remarkably, these weather conditions often come to serve to our advantage. Such was the case in an investigation in our following chapter, and such was the case here.

In this case, the cold windy weather outside allowed us to document several locations throughout the house with significant drafts caused by poorly-sealed old windows. We assume that's no longer the case now that renovations have been completed, but at the time we noted that the drafts could potentially account for some of the paranormal allegations. Indeed, a draft at the bottom of the main staircase was strong enough to blow papers around if not weighted down.

A commonly reported ghostly sighting at the Mansion is that of a ghost seen walking up the main staircase. Quite by accident, we found a way this could be explained by means of an optical illusion. The main room near the staircase featured a large chandelier hanging from the ceiling. At the bottom of the fixture was a large ball. If you walk from the east side of the Mansion into the main entrance room during low light conditions, that ball looks eerily like a person walking up the stairs. Of course that illusion wouldn't pass detailed examination, but it was quite convincing at a glance. People who thought they'd seen a ghost might easily have been scared into walking or running away instead of investigating closely.

Optical illusions caused by reflections and shadows of passing traffic, accentuated by imperfections in the antique glass, can also look, particularly when one is in a "ghostly" state of mind, quite a bit like shadowy people walking around the Mansion.

As usual, we're not going as far as to say these illusions account for all of the apparitions people have reported. But we do think they almost certainly account for at least some of the claims.

Just because our monitoring equipment didn't pick up anything unusual, though, doesn't mean nothing unusual happened during our investigations. Sometimes we see things independent of all our technological toys.

By far the most unusual thing that happened during any of these investigations—and right in the running for weirdest thing we've ever seen on any of our investigations—began during a seemingly quiet moment when most of our team were just watching the monitors while a few other people from the broadcasting crew who'd come with us were poking around the Mansion to see what they could find. After a while, one of the crewmen came back to our base of operations.

"You need to see what I just found," he said. He didn't elaborate.

"Okay, what've you got?"

"Just come and see."

Normally when someone insists on showing rather than telling, it's going to be something good, so a few of our team followed him downstairs where he pointed out a freezer. We opened it up and inside, we found...cats. Frozen cat carcasses. Maybe not *para*normal, but certainly not normal, either. Rarely are we at a loss for words, but that one stumped even us. We did know that Dr. Ikeler was a veterinarian and his family were the current residents at the time, so we figured they were part of his business somehow, but we didn't know exactly how or why the felines had come to their cryogenic end and the doctor wasn't around at the time to ask, so we had to just file it away for later.

A partial answer was given in an interview Dr. Ikeler gave for the *Castle Project* documentary. He admitted the cat freezer was his and explained that he'd wanted to preserve his two favorite cats in the freezer until he could construct a proper memorial and bury them with honor.[55] Seems weird to us, but fair enough. We understand loving one's pets. Trouble is, we'd seen the freezer for ourselves. We didn't count, but it contained more than just two cats. So we assume Ikeler's story is more or less accurate but that he just understated the number to make it sound more normal. To the extent any feline fridge could ever be considered normal. Or we suppose someone else could have come along and added to the frozen collection, but that notion raises so many more questions than it could ever hope to answer.

We know people reading these books are excited about the ghost stories, though, so we've saved one last unusual thing for last. Cats aside, it's the closest we came to documenting something legitimately anomalous at the Croke-Patterson Mansion.

On this investigation, we'd been brought in as the resident paranormal research group for an overnight radio show at the Mansion. After the evening broadcast, everyone was sent off to find a place to rest for the night before reconvening several hours later for the early morning broadcast. One of the radio station employees decided he was going to sleep in the basement. That had been Dr. Sudan's medical office and was at the time being used as a playroom for the resident children and he figured it would be well enough out of our way since we hadn't stationed any of our cameras down there for this particular investigation.

When he got to the room, he saw a person standing in the corner of the room by the fireplace. He assumed initially it was just one of the guests who hadn't cleared out when they were supposed to, but then he realized both that he couldn't think of how anyone could be unaccounted for in the building and, more disturbing, he could only see the person's head.

He ran back upstairs to announce what he'd seen. Several of us dropped everything we were doing to run and investigate, but by the time we got there, the figure was gone. We never found anyone else in the building. The next morning, one of the children said that was where they always saw a man standing in the corner.

Of course the correlation between what the radio employee saw and the child's claim is intriguing but far from dispositive. Regardless, the incident left us both excited and frustrated. After all these years, finally an opportunity to witness an apparition and we missed it by mere minutes!

So there you have it—arguably the most haunted house (now hotel) in Denver. Honestly we've only just scratched the surface of all the stories about this place, but hopefully that gives you a good overview. The stories are so eclectic and the building itself so spectacular both in its history and its architecture that

---

55   Higgins, B., Marcus, C., & Salvione, E. (Producers), & Higgins, B. (Directors). (2013). *The Castle Project* [DVD].

its one of our absolute favorites even though we weren't able to reach any final conclusions on its haunted status.

References & Further Reading

City and County of Denver (2023). Denver Individual Landmarks. City and County of Denver. <https://www.denvergov.org/files/assets/public/v/4/community-planning-and-development/documents/landmark-preservation/individual_landmarks_list.pdf>

Denver Post (2009). Rocky Mountain News history timeline. *The Denver Post*. <https://www.denverpost.com/2009/02/26/rocky-mountain-news-history-timeline>

Emrick, C. (1973). Croke-Patterson-Campbell Mansion: National Register of Historic Places Inventory – Nomination Form. [Register No. 73000467]. *National Register of Historic Places*. <https://catalog.archives.gov/id/84129551>

Friends of Red Rock Canyon (n.d.). Red Rock Canyon Stone in Denver's Historic Buildings. *Red Rock Canyon Open Space*. <https://redrockcanyonopenspace.org/education/history/red-rock-canyon-stone-in-denvers-historic-buildings/>

Goodstein, P. (1996). *The Ghosts of Denver: Capitol Hill*. Denver, CO: New Social Publications

Higgins, B., Marcus, C., & Salvione, E. (Producers), & Higgins, B. (Directors). (2013). *The Castle Project* [DVD]

Ikeler, D. (2008). *Life Extension: How to Use Your Neurophysiology to Live 150 Years!* [self-published].

Lamb, K. (2016). *Ghosthunting Colorado*. Covington, KY: Clerisy Press.

Leggett, A. A. & Leggett, J. A. (2011). *Haunted America: A Haunted History of Denver's Croke-Patterson Mansion*. Charleston, SC: The History Press

Patterson Inn (n.d.). About the Inn. *Patterson Inn*. <https://www.pattersoninn.com/about/>

Peterson, E. (2023). Only in Denver: City Ditch, the 'Oldest Working Thing.' *Visit Denver*. <https://www.denver.org/blog/post/city-ditch/>

Wier, J. & Miller, J. (1997). Of Things Medical in Middle Park. *Grand County Historical Association Journal, 14*(1)

Zimmer, A. (2016). Denver's Most Haunted House. *Colorado Virtual Library*. <https://www.coloradovirtuallibrary.org/resource-sharing/state-pubs-blog/denvers-most-haunted-house/>

Zimmer, A (n.d.). Croke-Patterson-Campbell Mansion. *Colorado Encyclopedia*. <https://coloradoencyclopedia.org/article/croke-patterson-campbell-mansion>

14

I'm Not Here: The Brook Forest Inn

Whenever we tell one of our stories, we invariably call it one of our favorites. That's because, each in their own way, every case contains something fascinating and we love telling the tales. But the Brook Forest Inn, located at 8136 South Brook Forest Road in Evergreen, Colorado, stands out as a favorite among favorites.

The Inn itself is a beautiful building with a delightful history that sits at the heart of a replicated Swiss village nestled in the Rocky Mountains. But it's our investigative activities, which lasted some fifteen years, that earned the Brook Forest a place of honor as one of our favorite haunts. In both senses of the word.

Figure 14.1. The Brook Forest Inn. Photo: Bryan Bonner.

The History

When we set out to start investigating paranormal claims, we already had a passing appreciation for history but did not expect that our chosen vocation would put us on a path toward becoming de facto local historians. Yet that's become one of our favorite parts of our work. There's something both magical and important about digging up those key little facts that help us remember our collective forebears. Few things are more rewarding than being able to tell those stories to others. Teachers have come to love us because we rope their students in with the promise of a good ghost story—which we'd like to think we deliver well—but along the way we "trick" them into learning some real history.

Nowhere has our position as inadvertent historians been more prominent than in the case of the Brook Forest Inn. Since we spent over a decade working on our investigation, word eventually got out that we were collecting stories from its history. Over the years, we've received phone calls from individuals, quite literally on their deathbeds, who wanted nothing more than to spend an hour or two telling us their stories from when they used to work at or visit the Inn. Not even just ghost stories, though we've collected a few of those, but just stories from the history. Even as we were preparing this manuscript, we received another such letter. That's quite an emotional experience and preserving those stories is a responsibility we take quite seriously.

Outside of paranormal lore, relatively little has been written about the Brook Forest's history. Of course we've collected a few citations here and there, but a lot of what we know about the place came from those first-person interviews with Inn visitors and erstwhile employees.

The story of the Brook Forest Inn begins with Edwin F. Welz (1880-1956). Born in Vienna, he trained as a chef[1] and worked in that capacity at a variety of fine hotels throughout Europe. His first journey to the United States occurred in 1903 when he arrived in Boston with no money and no command of English, though he was able to find his way working at a Boston hotel for $30 per month (equivalent to a little over $1,000 in 2024) while studying English at night. By all accounts, he earned a reputation as not only a hard worker but a world class chef during this period and took a variety of positions cooking at not only high-class dining establishments but for politicians and foreign dignitaries.[2]

In 1907, he returned to his native Europe and purchased shares in the Pension Waldeck Hotel in Langenbruck, Switzerland, where he met and fell in love with Marie A. Jenny (1884-1969). But the opportunities of the American Dream were too much for Welz to resist so he returned to the country, this time with Marie, where they married in New York in or around 1910 then moved to Denver where they applied for citizenship while working at a variety of restaurants and hotels, including the Brown Palace (whose ghost stories will be included in a future volume of this series). They had their only child, Edwin H. Welz, in 1911,

---

1   We like him already!

2   Atencio-Church, S. & Atencio, B. (2009). Brook Forest Inn: National Register of Historic Places Inventory – Nomination Form. [Register No. 09000567]. *National Register of Historic Places*. <https://catalog.archives.gov/id/84131063> (accessed July 13, 2024).

then moved to what would later become Evergreen in 1913, taking over an abandoned homestead claim of some 160 acres of mountain land, later expanding to 400 acres by 1919.[3]

The Welzes loved their new home in America but they also remembered their European roots. Their goal was to create a little replica of a Swiss village right in the middle of Colorado. And we'll be damned if they didn't pull it off! Beginning by making small improvements to the land while Mr. Welz continued commuting to his restaurant jobs in Denver, they ended up creating a little village with nine chalets, a castle tower of quartz stone, and the flagship Brook Forest Inn which could accommodate up to 100 guests at any given time. The Inn officially opened for business on May 17, 1919. In 1921, Mr. Welz became the first postmaster of the newly established Post Office in what was then called Brook Forest and is now part of Evergreen.[4] He was even able to convince the United States Forest Service to create a trail leading to a waterfall on public land which still exists today.

The Inn itself reached the height of its popularity in the 1940s and 1950s, offering not only accommodations but entertainments and Mr. and Mrs. Welz's fine cuisine. It was designed in the Swiss Chalet style with some Tudor Revival and Rustic influences, as were much of the village the Welzes created, and operated almost continuously from its opening in 1919 until quite recently (as of this writing, it has sold to a new owner and remains closed; we're not yet sure when it will reopen or how it will operate at that time).[5]

What was life at the Inn like? We know from historic records that the hotel offered fine enough amenities to attract the likes of Molly Brown, President Franklin Delano Roosevelt, and Liberace. But at the same time, the Welzes and their staff were welcoming and hospitable to everyone and had a particular fondness for children and families, a fondness which grew after the untimely death of their only son in 1922. Food was prepared by both of the Welzes, who raised their own livestock and grew their own vegetables.[6]

People who used to work at the Inn have given us a glimpse into life at the Brook Forest and the character of the Welzes you wouldn't necessarily find in dry historic records. One individual met the Welz family when her grandparents used to live next to the Welzes. At that time, Mr. Welz was still commuting to work at Bauer's restaurant and bakery in Denver during the week and joining his wife at the Brook Forest on weekends. When she was nineteen years old, this family acquaintance took a job, along with her sixteen year old brother, at the Inn, where she worked for four years. In her own words, Mr. and Mrs. Welz were both "firm and exacting" as well as "kind and fair" to the staff. She painted a picture of kindly and generous people whose only harshness (if one dares to use the word)

3   *Ibid.*

4   *Ibid.*

5   History Colorado (n.d.). Brook Forest Inn. *History Colorado.* <https://www.historycolorado.org/location/brook-forest-inn> (accessed July 13, 2024).

6   Atencio-Church, S. & Atencio, B. (2009). Brook Forest Inn: National Register of Historic Places Inventory – Nomination Form. [Register No. 09000567]. *National Register of Historic Places.* <https://catalog.archives.gov/id/84131063> (accessed July 13, 2024).

was their insistence on the highest quality of work and perfect cleanliness of the Inn—on one occasion Mr. Welz ran his finger along the top of a door frame and reprimanded her because he'd found dust. She said the staff only ever called Mr. Welz "Mister Welz," but came to refer to his wife as "Ma." Apparently they had a particular fondness for her brother because he was about the same age as their late son.

Mr. Welz was also quite protective of his staff and his firmness came to their defense when necessary. On one occasion, a new cook at the Inn took a certain fondness for the same young woman who shared her remembrances of the Welzes' character. His feelings were not reciprocated. One day, he cornered her as she attempted to cross a bridge and wouldn't let her pass unless she agreed to go on a date with him. Mr. Welz observed the encounter from the Inn and called the girl into his office. He asked if she didn't like the new cook. She said she didn't dislike him but just didn't want to date him.

"I see," was all Mr. Welz said. He put the young man on the bus back to Denver the next day.

Staff always ate before the guests and were expected to help entertain the guests if necessary. Whenever someone needed another player for their bridge or tennis game, employees would fill in. Mr. Welz himself was known to take the time to play chess with his guests and staff. Employees and guests alike developed a strong sense of community, often riding horses into Evergreen to attend a dance or see a movie together. At Christmastime, the staff got to spend their holiday at the Inn as guests rather than employees, with the Welzes cooking and entertaining.

Staff were heroically protective of the property, both during the Welz era and after the Welzes had sold it. In 1964, a fire broke out, likely caused by a malfunctioning boiler, and the staff evacuated the building and fought the blaze with fire extinguishers until the Evergreen Volunteer Fire Department arrived on scene.[7]

Through the years, the Inn and surrounding properties have passed through a lot of hands. In 1946, it was acquired by Christian and Jenne Maurer, though the Welzes took back over in 1952 and held onto it until Mr. Welz died in 1956. At that time, Mrs. Welz sold the property back to the Maurers and returned to Denver where she spent the rest of her life until she died at David Nursing Home in 1969. Her ashes were scattered near the Brook Forest Inn. In the years since, the Inn has continued much as it did during the Welz years, offering fine accommodations as a lovely mountain hotel, though it has also been a lounge, bar, concert venue, and during a portion of our own time at the Inn it was operating as a Microsoft certification school (about which more shortly).

Paranormal Claims

Both staff and guests have reported a variety of paranormal claims—mostly in the form of ghost stories—at the Brook Forest Inn. Many of them are hard to trace to their origin, seemingly being passed from one owner to the next or re-

---

7   Canyon Courier (November 19, 1964). Cause and Effect. *Canyon Courier, 10*(47).

peated with slight variations from one newspaper article or blog post to another. Nevertheless, we've done our best to catalogue the tales as best as we can.

Any hotel needs a ghost. And any haunted hotel worth its salt has the ghost of a young child (or children) often claimed to run up and down the halls, frequently playing with a ball. One previous owner said she was inspecting the Inn. While walking up the main staircase, her son moved to the side of the stairs for no reason. When she asked what he was doing, he said he was moving out of the way of the little boy. Of course the mother didn't see anyone and there were no other children present in the Inn at the time.

The tales of this young boy have been connected specifically to the Welz family. The lore suggests the ghost was that of the original owners' son (Edwin H. Welz) who died at a young age of a lung disease (often identified in articles as influenza). A particular room on the third floor is supposed to be the lad's former bedroom and the most common place to see his ghost.

Arguably the most prominent ghost story of the Brook Forest Inn centers on "Carl." Details of the story vary in their telling. Depending on who you ask, Carl was either an Inn employee, a stable hand from the nearby livery stable or perhaps just a guest. Whatever his relationship to the Inn itself, he was also supposed to be either the boyfriend or fiancé of one of the Inn's chambermaids. Legend has it that, probably distraught over a breakup (though his motivation is unclear) Carl murdered his girlfriend/fiancé in one of the bedrooms. Most versions of the story have this murder being done by strangulation. Distraught over the guilt, he then hanged himself either in his bedroom, the small house next to the Inn, or the livery stable. An alternate version omits the murder and has Carl dying of a heart attack in his bedroom. Despite the variances in the details, Carl's ghost is the most commonly reported at the Inn.

One claim is that if any male dares to sleep in the bedroom where the murder took place, he'll feel either ghostly hands around his neck or pressure from an unseen force on his chest, making it difficult to breathe. Perhaps that's the ghost of Carl's girlfriend?

Carl's own ghost has been seen throughout the Inn, with most of the stories focusing on the three candidate locations for his alleged suicide.

The creepiest story (we think) from the Inn might or might not have involved Carl's ghost. When it was relayed to us by a former employee, he identified the spirit as Carl but nothing about the story necessarily limits us to that interpretation. Regardless of the ghost's identity, here's what happened. One day the manager at the time was on his way into the kitchen along with the former employee who told us the story.

When they stepped into the kitchen, they noticed a man standing at the far end of the room near the coffee bar (where the meat slicer is in our photo taken during one of our investigations). Because they were just opening up for the morning, no one was supposed to be there.

"Can we help you?" one of them asked.

Without turning around, the man replied, "That's okay. I'm not here."

He then walked out of the kitchen and into the small office at the far end of the kitchen. It's important to note that room has only one door in or out. When the manager and his employee tried to follow the mysterious man into the office to inquire further, they found the room empty. Somehow, he'd just vanished!

The reason our contact believed this might have been the ghost of Carl was that he said whenever someone saw Carl's ghost, they could never see his face. According to this witness, Carl was about five feet, nine inches tall, of medium build with black hair, always seen wearing a gray suit. The mysterious man in the kitchen matched this description. And of course he never turned around so they never saw his face.

Figure 14.2. The kitchen where the ghost was seen. Photo: Bryan Bonner.

The same witness also said odd things would happen in Carl's room. Doors would open and close themselves. Mysterious sounds would be heard. And if a pitcher of water were ever left in the room, it would be partially empty by the time the guests returned. Not necessarily connected to Carl's ghost, he also said he frequently heard sounds coming from the third floor at night, including the sounds of people walking around and having conversations when no one was present.

A variety of other ghostly claims have not been attributed to any particular spirit or historic individual. One report claims a full apparition has been seen in a room called the Monte Carlo II suite, on or near the balcony. Others claim to have seen people walking around or in the building even when it's been vacant. Some guests have complained of the smell of rotting fish on the second floor with no apparent source or cause ever having been found.

Paranormal investigators and ghost hunters (other than ourselves) have add-ed to the lore in their own right. One such investigator set out to explore the Inn, set up all of his equipment (we're not sure how much he might have used but if he was anything like us, you already know it was a lot of gear), then promptly exited the building and refused to return. Someone else had to go back to pick up all his things.

Similar rumors claim a paranormal investigator killed himself because of the hauntings at the Brook Forest Inn.

We don't spoil much about our investigation section below to point out that we haven't verified either of these stories. The latter was actually once attributed to Rocky Mountain Paranormal, but we can assure you, at the very least, that never happened to us or anyone of our acquaintance.

A psychic brought in to investigate the Inn claimed the ghosts were either exclusively or primarily past employees who haunted the building because they were upset about their treatment while they worked at the Inn. Of course that doesn't seem to be at all in character with what we've heard from past acquaintances and employees of the Welzes.

According to the individual who wrote to us just as we were preparing this manuscript (an individual claiming to be the child of a former owner), the Inn is haunted by children who run in the halls, a maid who wanders the second floor, and more. This individual claims to have heard ghostly sounds on the third floor, and to have been frightened by the wine cellar as a youth. Another possible ghost, we're told, could be the brother of one of the partners who owned the Inn at the time. Allegedly he was "ran over by a road grader" near the swimming pool. Because that letter only reached us literally days before sending this book to the printer, we haven't been able to look into any of those claims in detail yet.

Finally, the most unfortunate of all the stories and rumors about the Brook Forest isn't actually a ghost story. It's not even a supernatural claim, though we do consider it "paranormal" in the sense that it's so extraordinary as to be outside of normal experience.

According to the rumor, the Brook Forest Inn was once a favorite spot for German Spies during World War II. The Nazis, the legend goes, would pose as bicycling tourists but were actually making maps for Germany. More extreme versions of the story even hold that the Inn was used as a Nazi internment camp to hold American prisoners of war, and that "Nazi gold" is buried on the property. As evidence of this, people point to various swastikas incorporated into the building's décor.

Our Investigation

We were able to investigate at the Brook Forest Inn for a long time. Often our investigations are limited to just one or two evenings. This one lasted some decade and a half of on-and-off investigation. So we have plenty of stories.

When we were first contacted by the then-owners of the Inn, it was functioning as a Microsoft certification school. At the time, computer certifications weren't done the way they're done these days. Today, when you need a certification, you sign up for an online course, watch video lectures, take quizzes, and download your certificate. Or if you're willing to be a bit shadier, you give your log-in credentials to someone else and have them earn a certification in your name. But at the time, certification school was more akin to boot camp. Students would show up to the Brook Forest Inn for a six-week certification program. During that time, they'd take their classes in a computer room set up in the hotel, eat at the hotel restaurant, and sleep in bedrooms upstairs.

Figure 14.3. The Brook Forest Inn was once a computer school. Photo: Bryan Bonner.

It was an efficient operation. It was also great for us. The schedule of the certification school meant that when school was in session, six weeks at a time, the place would be bustling. Between sessions, though, the Inn would be empty but for a single caretaker for months at a time. For our part, that meant we knew we had a place we could go, almost any time we wanted, for the majority of the year. Whenever we wanted to test a new instrument or technique, try out a potential new recruit, or even just got bored and wanted to go look for ghosts, we could just head up to the Brook Forest. For his part, the caretaker also loved us. Once we'd established ourselves as known and trusted people, he knew whenever we were there, he had a free ticket to take the night off and go somewhere else because the Inn was going to be in good hands. As a result, we spent a *lot* of time examining every corner of that building.

That's the way paranormal investigation really ought to be done. Spend enough time in a location and you not only maximize your chances of experiencing any anomalous phenomena that occur at irregular intervals, but you also become intimately familiar with the building itself and its various quirks. And it allows for plenty of follow-up investigation into anything strange that does happen. It's only a matter of pure cost and convenience that keeps us (and others) from conducting every investigation in this manner.

Before we get to our on-site visits, though, we need to dive in to a bit of background research.

First of all, there is no evidence at all—we repeat, none—that there was ever a Nazi presence at the Brook Forest Inn. We believe the rumor started because there are indeed swastika patterns in the building's décor.

However, swastikas were not a Nazi invention. Though it's now recognized—and shunned—worldwide as perhaps the most recognizable symbol of Nazi oppression, the design of the cross with bent arms predates its German use by centuries and is recognized in Hindu, Buddist, Jainist, and Native American religious symbology, and also pops up from time to time in pre-Christian Europe.[8] Likely the architect included those designs in the Brook Forest Inn either because it's a relatively simple geometric design or because of its association in various religions with notions of good fortune.

The timing also doesn't work out. The Welzes immigrated to the United States and built the Brook Forest Inn long before even the founding of the Nazi Party in 1920 and certainly before the swastika's use in the German flag from 1935-1945. Furthermore, any association with the Nazi Party would have been incredibly out of character for the Welzes, who even hosted President Roosevelt at their Inn (!), nor is there any evidence of a Nazi presence in the Evergreen region during the war. And as for the rumors of buried Nazi gold, all we can really say is no one ever found any gold buried on the property, and the building has passed through the hands of enough owners that surely someone would have found it if it ever existed.

That story in particular bothers us not only because it's entirely false but because it besmirches the reputation of people who were apparently so kind to their employees and staff that we still have people calling us from their deathbeds to share their favorite stories of the Welzes and the Brook Forest Inn.

On the other hand, we were able to partially confirm one of the other stories. Recall that one of the ghostly claims was that the hotel is haunted by the ghost of the first owners' young son who died of a lung disease, possibly influenza. After more hours than we care to admit digging through microfiche newspaper archives in various libraries, we were able to find two obituaries from separate newspapers.

The first simply identified Edwin H. Welz, aged ten years, as having died on March 25, 1922 and provided details for a memorial service. The second went into a bit more detail: "Edwin Welz, only son of Mr. and Mrs. Welz, of Brook Forest, passed away at his house Saturday morning, pneumonia being the cause. Mr. and Mrs. Welz have the sympathy of all their friends."

Does that prove the ghost story? Certainly not. But it does add some degree of credence to the story in the sense that it's another piece of the puzzle. We can't say whether young Edwin haunts the Brook Forest Inn, but we can at least say that the ghost stories largely got the details of his life and death correct. Yes, the lore tends to blame influenza and the obituary blames pneumonia, but to that we can respond either by hypothesizing that influenza may have caused the pneumonia (we can't confirm that to be the case here but we do known influenza to be a common cause of pneumonia) or simply pointing out that minor details often get slightly altered in a story's retelling but the fundamental truth of the history remains intact.

To actually prove the ghost story, we'll need a lot more evidence. But paranormal investigation isn't the kind of work where we quickly find easy answers.

---

8   Chandra, S. (1998). *Encyclopedia of Hindu Gods and Goddesses*. New Delhi, India: Sarup & Sons.

We're assembling puzzle pieces one by one. Had we confirmed that Edwin had died elsewhere, at a different age, or of a fundamentally different cause, that would have largely discredited the lore. The answers we did find don't prove it but they do at least leave it within the realm of historical plausibility.

Other stories, like the "I'm not here" incident in the kitchen, are impossible to research. We simply weren't there when it happened. Despite attempts to recreate the scene with video monitoring in place, it never recurred in our presence. We did interview the witness and found him credible, but that's the kind of one-off story that gets us really excited and frustrated all at the same time because there's nothing we can do to prove or disprove it.

A common challenge when we're investigating places as large as the Brook Forest Inn is that we don't have the equipment or manpower to monitor the entire place at once. Our long-term arrangement at the Brook Forest meant we could actually check out the entire place, albeit piecewise rather than all in one evening. Over the years, we conducted extensive monitoring of the bar, the breakfast room, the dining room, all rooms on the second floor, the corner room and storage room on the third floor, the main staircase, the second and third floor hallways, the balconies, the kitchen, and the main entrance, as well as minor or less formal monitoring of the rest of the property.

Many of our trips up to the Brook Forest didn't amount to much. Always enjoyable, they weren't always terribly interesting in the paranormal sense. Of course we always employed our usual array of video, audio, EMF, seismic, and other forms of monitoring and measurement, but often didn't get any promising results. Rather than boring the reader by walking through over a decade's worth of null results only occasionally punctuated by interesting happenings, we'll simply highlight the parts where something did happen.

One evening, a team member of ours started to get tired and wanted to rest. We've mentioned before that we have a protocol for such situations. Team members are absolutely allowed to sleep on the job, but only on the condition that they have to be "bait." This time, we sent the sleepy investigator to the room in which Carl supposedly murdered his girlfriend—the same one where male visitors are sometimes supposed to feel like they're being choked by a ghost. Some men would probably find that an enticing prospect, but our team member just wanted to sleep so he set up video and audio monitoring in the room (we always watch what happens to our "bait" team members) and went to sleep.

After about an hour of sleep, he started stirring and whimpering. Eventually he woke up and claimed he'd heard sounds of heavy breathing "right in front of my face." While monitoring, we thought the sounds were his own troubled breathing, but he insists he heard it coming from just in front of him. Obviously we can't translate audio into print, but it really does sound like a deep, almost raspy, heavy breathing. We frequently play the recording for audiences when we discuss this case in our public lectures.

Another incident perhaps related to Carl the ghost happened in a different room. We're not sure which room Carl killed himself in. As a matter of fact, we're not even sure if Carl was a real person as we haven't yet found historic records to corroborate the story. But one of the rooms he might have killed himself in is a bedroom on the Inn's second floor.

On a different investigation, we'd established our base of operations in the hallway just outside that room, with monitors in the room itself. After several hours of quiet, one of our team members happened to look up and notice the ceiling fan in "Carl's room" was slowly spinning. Not that it was turned on; rather, it was just gently rotating. He pointed it out to the rest of the team because he was all but certain the fan had been turned off hours before.

Skeptical as we always are, we went through a variety of different possible explanations. Perhaps it was still winding down from when it had been on earlier. To test that, we turned the fan back on, let it get up to full speed, then turned it off and timed how long it took to run down and stop. It stopped spinning after about sixty seconds or so. We might not have kept a record of exactly when we'd turned the fan off, but we knew full well it had been longer than that, so clearly it wasn't still just winding down.

Another thought was perhaps a draft blew the fan. But when we stopped the fan, we kept it under video surveillance for the rest of the night and it didn't start back up, as we would expect if a breeze was blowing it.

Of course it's possible a momentary draft might have made it spin just that one time. We can't definitively rule that out. But neither have we been able, despite a lot of subsequent searching, to identify any draft that causes the fan to blow in the breeze. We've checked plenty of times since then and that fan hasn't behaved the same way since.

Visual manifestations are always popular among paranormal enthusiasts. Our experience, though, has been that frequently we *hear* more unusual things than we *see*. Such has been our experience at the Brook Forest Inn. With surprising regularity, we'd return from an investigation and review our audio to find it full of unusual sounds: footsteps, doors opening and closing, and an assortment of other pops, thumps, and creaks.[9]

Once again, we're not saying these are proof positive of a haunting. We don't rightly know what the sounds are. Old buildings often make noise, particularly as the weather changes. Pops, thumps, creaks, and all manner of noises are common and before we can even claim "anomaly" (much less "ghost") we have to rule those explanations out. However, we have spent enough time in the Brook Forest to be reasonably familiar with it's normal "house sounds" and we've been careful to document everyone present during our investigation. Some of the sounds we've recorded sound eerily like footsteps and doors closing and we're quite certain no one was walking around or closing doors when they were recorded. We may not know exactly what they were, but we do at least know what they weren't.

And then there's "the" night at the Brook Forest. The one that made our mental gears grind so hard we're surprised smoke didn't start pouring from our ears.

We arrived in the late afternoon or early evening of what was to become

---

9   A collection of such sounds, assembled from audio recorded during approximately eight hours of investigation, is included in a news broadcast of one of our investigations and, as of this writing, is available for viewing and listening at https://www.youtube.com/watch?v=r4LaH496jBw.

quite a wintry night. As soon as he saw us, the caretaker said something very much to the effect of "Thank God you guys are here. Take the keys. I'm going to the pub."

And so he did. He went off to enjoy himself on a night off and we set up for a nice evening in the Brook Forest with no one else around. We'd selected the corner room on the third floor as our base location. This was the room Edwin H. Welz was meant to have died in back in 1922, so we were quite acquainted with the room after many other investigations. From the room, we were monitoring the cameras and microphones we'd placed in the hallways and various common areas, including the main entrance and bar and restaurant area downstairs.

At about 2:30 a.m., we heard the unmistakable sound of the front door opening and a large group of chattering voices entering the bar. We were livid! Given the time of night, we assumed the bar had closed down and the caretaker had brought a crowd back with him to continue the party, right in the middle of our investigation.

Two of our people went downstairs to investigate. No one was there. Everything was just as we left it and we didn't see or hear a soul. A bit confused, we checked outside. We've mentioned before that for whatever reason we tend to do investigations on blizzardy nights. This time, we were grateful for the snow because it allowed us to check for tracks. Nothing. The caretaker's car wasn't out there, there were no tire tracks on the driveway, and no footprints in the snow. To do our due diligence, we even checked the caretaker's "apartment" room. Nothing and no one. He hadn't come back at all.

Even more confused, our two people went back upstairs to tell the others what they'd found. As soon as they opened the door, the people inside said, "What happened? The sound just stopped."

"What do you mean, it *just* stopped?"

"It just stopped a couple seconds before you got back."

Figure 14.4. Did a ghostly party occupy this bar? Photo: Bryan Bonner.

The entire time two of our people were wandering around downstairs in an empty and quiet Inn, the rest of our people were upstairs listening to audio of what sounded exactly like a party—complete with multiple distinct voices—recorded on microphones in the very same rooms and hallways the first two people could confirm were empty and silent. Somehow we managed to record a party on our microphones that people in the room couldn't see or hear.

To describe the audio since you can't hear it, the relevant portion of the recording begins with what sounds exactly like the front door opening (we've recorded the door opening and closing on other occasions and it really does sound the same as in the anomalous audio). Immediately, we then hear multiple voices. No one can tell what any of them are saying. It sounds exactly like the indistinct chatter of numerous simultaneous conversations happening in a bar or restaurant.

We've never been able to replicate or explain that event. Audio engineers have checked out our recording and as far as they can tell it's just a recording of a bustling party. We have no idea.

Even at that, we're not ready to say we've proved a ghost story. That recording is mighty curious, but all we've proved is that something unusual happened that we can't explain. To go as far as to say we've proved a ghost, we'll need more direct and specific evidence. Beyond any doubt, though, that's one of our most intriguing experiences.

We continued investigating for several years after the computer training school moved out. Eventually, though, all good things must come to an end. One subsequent owner (not the current owner) decided they didn't want anything to do with the ghost stories, and we lost our access to our favorite long-term investigation site. Hopefully subsequent owners will let us back in because we'd love to be able to follow up on some of those mysterious, intriguing, fascinating, and frustrating occurrences. For now, though, we leave the case of the Brook Forest Inn open and unsolved. Spectacularly so.

References & Further Reading

Atencio-Church, S. & Atencio, B. (2009). Brook Forest Inn: National Register of Historic Places Inventory – Nomination Form. [Register No. 09000567]. *National Register of Historic Places.* <https://catalog.archives.gov/id/84131063>

Canyon Courier (November 19, 1964). Cause and Effect. *Canyon Courier, 10*(47).

Chandra, S. (1998). *Encyclopedia of Hindu Gods and Goddesses.* New Delhi, India: Sarup & Sons.

History Colorado (n.d.). Brook Forest Inn. *History Colorado.* <https://www.historycolorado.org/location/brook-forest-inn>

PART TWO:
Private Residences

Though the public venues we've discussed so far (and will continue to discuss in future volumes of this series) are arguably of the greatest general interest, they don't represent the entirety (perhaps not even the bulk, though it varies from year to year) of our work. On a regular basis, people write to us or approach us at one of our public lectures and ask for our assistance with their personal paranormal experiences.

Private cases such as these come in all shapes and sizes. Some of them never proceed past some initial email discussion or perhaps a quick forensic analysis of a single photograph, while others may involve substantial on-site research on par with the kind of work we've done in some of the public cases.

For this second part of the book, we'd like to take you through a tour of some of these private cases, but there are a couple of caveats to get out of the way before we do so.

First, while we're attempting to provide a complete record of our case files in these books (though it will take several volumes to do so, especially since we're still actively conducting research on new cases), we're not going to document every communication we ever receive. Many cases simply don't go anywhere, and while we maintain records of these cases in our private files, we wouldn't presume to bore the reader with repetitive details of cases we were never able to fully investigate. On the other hand, neither are we presenting only the most outstanding and extraordinary cases. We do want to provide a reasonable overview of our activities.

Second and most importantly, because these are not public venues or public figures, we will not identify the individuals into whose homes and lives we've been invited. Most of the time, we can accomplish this simply by omitting their names or assigning pseudonyms. Every once in a while, some feature of the case narrative itself might be potentially identifying. When that happens, we will disguise the individuals' identities in a variety of ways, but never in such a way as to alter the relevant data we're trying to present.

With regard to photographic evidence, we have similar concerns. Occasionally, we may have a photograph that's restricted enough in its contents (or can be cropped to be so restricted) that it can illustrate the relevant scene without risk of identifying or embarrassing the client(s). When that's the case (in this volume as well as throughout the series), we'll use genuine photographs. The rest of the time, we'll use artists' renditions to illustrate the cases.

On occasion, we've mentioned our work to people of our acquaintance who aren't involved in the same kind of research. Reactions to the claims made by our clients are not always positive. It's easy to dismiss some of these claims as being those of crazy people. But that's not been our experience. While there certainly are some people who suffer from a variety of mental health conditions ("crazy" is not a technical term) and occasionally they cross our path, most of the people who reach out to us are fundamentally sane and normal. Paranormal lore is a part of our culture and plenty of reasonable people believe in it.

Often, perfectly sane and normal people who simply happen to have a belief or interest in paranormal phenomena have some unexplained event in their lives, and we consider it our duty to help them understand what they're dealing with. If we can find a natural explanation so they don't have to be afraid of their own homes anymore, we consider it a good day's work. On the other hand, if we could document a legitimate paranormal event to validate what they thought they were experiencing, we'd also consider that a good day's work.

In the pages to follow, you'll meet a diverse cast of characters, and we hope you enjoy their paranormal stories and our research into the same.

Note: though the chapters in Part One were divided into sections for history, paranormal claims, our investigation, and further reading, the private cases will not be. Most of these residences don't have much history to speak of, and further reading on these particular cases is typically nonexistent. Instead, we'll present the entire story in narrative form.

ESSAY

The Importance of Choosing the Right Paranormal Investigator

There's something important we need to get off our collective chests, and this seems like a good place to do it, though we also need to tread carefully because we're about to criticize some of our colleagues in both the paranormal and skeptical worlds. It's important for the reader to realize that we're not trying to put anyone down and we're not trying to insist our way is the only legitimate way to do things. However, we are of the opinion that not all paranormal investigation teams are equal and while there is plenty of room for difference in philosophy or methodology while still remaining within the bounds of legitimacy, there are some groups and/or practices that we think have, at least on occasion, crossed some lines that oughtn't to be crossed. As such, we decided to open this section of our second volume with some tips for those who might be thinking about reaching out to a paranormal investigator for assistance with some unexplained phenomenon.

To be a bit blunt right at the beginning: we really like it when we're the first team someone contacts. That's not ego speaking (or at least not *mere* ego), and we'll explain our reasoning in more detail. But, while we're always happy to assist people even if they've worked with others in the past, we've had to add the question of whether an individual has contacted another team to our initial interview process because we've found that in many—not all, but many—cases in which someone has already had someone else out, that other group's perspective has colored the client's perception of what's happening, and possibly even contaminated the scene.

One has to realize that the operative word in "paranormal investigation" is, in fact, the second word. We're not here to force someone into believing as we do about ghosts, aliens, religion, the nature of the universe, or anything else. Our job, purely and simply, is to do our best to determine, through as rigorous of methods as possible, what might or might not be happening. That means that when someone experiences something unusual, we don't have the luxury of assuming it to be whatever entity we might think or hope it could be. Neutrality is key, as we've explained both in our introductory remarks at the beginning of each volume in this series and throughout the cases. The trouble occurs when other teams do not employ the same kind of methodological (or even better, philosophical) neutrality.

Some examples are in order. These cases—at least the ones that have progressed to a conclusion—will be addressed in their own case files throughout this series (the one in Chapter 21 of this volume is a doozy!), but we'll highlight a few key points from cases in which we were not the first team called.

The most recent such case (and one that actually won't likely be featured in a future volume of this series because the client stopped responding to our letters before we could finish the investigation) is both a good example and the one that prompted us to include this essay in our book. We were contacted by an individual who said she was referred to us by another investigator who "confirmed the worst" and told her she had a demon in her house. She was naturally terrified, and we remain concerned about this client's welfare since she did stop responding to our letters after a brief exchange in which we asked for more information. The details of the case aren't important here except to point out that the initial phenomena she claimed she was experiencing in her home were, while certainly unusual, not the stuff of horror movies. They were what we'd consider fairly run of the mill paranormal claims. But things got out of hand when she reached out to this other investigator—whose work was already known to us from past encounters—and he escalated her anxieties instead of trying to resolve them.

One client contacted us after another paranormal team, instead of neutrally investigating, took the opportunity to mock the client's culture and religious beliefs.

A case that will appear in a future volume involved a family that eschewed medical advice in favor of a religious interpretation of their child's illness, to the extent that they would only work with people who confirmed their preconceived notions about the case, which ultimately led to irreversible harm to a child and prison terms for neglect.

Many clients for whom we were not the first team have reported not only the initial phenomena that started the whole process, but the interpretations supplied by prior investigators. While this may not involve the kinds of physical or psychological harm as some of the above examples, it does make our job harder because we have to be all the more careful to separate the initial claim from the interpretation that's been mixed in with it. It can also create methodological expectations on the part of the clients, who might assume our methods will be the same as the other teams' when in reality they may be quite different. Indeed, we even lost one case because a client was under the impression, based on prior experience with other groups, that a paranormal investigative outfit must necessarily employ a psychic (something we do not do for reasons we've explained elsewhere).

Another case—one you've already encountered if you read our first volume—involved a terminally ill patient who we were eventually able to determine simply wanted someone to talk to and (probably unconsciously) invented paranormal excuses to reach out to someone. As it happens we were the first ones on the scene in that case, and we're grateful that we were because we imagine that could have gone very wrong if someone confirmed the paranormal stories for this person instead of correctly diagnosing the situation.

Hopefully those examples help to illustrate why we like to be the first on scene. Not that there aren't other groups who can do things correctly (even if perhaps somewhat different from our own approach), but we simply can't trust that all other groups will hold themselves to the same standards that we and other reputable organizations do.

Paranormal investigation doesn't have any kind of meaningful certification process.[1] As such, *caveat emptor* is the order of the day. Still, one shouldn't avoid all contact with paranormal investigators. There are legitimate and reputable people out there who specialize in helping people understand and cope with anomalous experiences. The trick is in knowing which one you ought to work with. And it's not an easy decision to make, but we have come up with a few guidelines we think will help people.

1) If at all possible, work with Rocky Mountain Paranormal first. We're only half kidding. If you're in our region and need some help, we'd like to think we're the best game in town and we're always happy to work with new clients. So if you've been enjoying these books, we're happy to work with you and we'll give you the same dedicated treatment we give everyone else. But to be more serious, the remainder of our recommendations will be more general.

2) Don't hire a paranormal investigator who charges a fee beyond actual expenses or room and board. As we discussed in our essay on ethics, we don't think it's appropriate to charge people for services related to admittedly fringe science. Actual expenses, we understand. Things can get expensive. But even at that, we believe paranormal investigators should do their absolute best to self-fund. And if they are asking you to contribute to expenses, make sure it's all negotiated in advance and you know exactly what you're paying for.

3) Look for someone with a lot of experience in the field. Of course new people can still be legitimate, so this isn't a hard and fast rule. But something we've discovered is that it's very easy for someone to want to be a paranormal investigator. Setting up an organization and conducting a first investigation is child's play. Sticking with it for years—that takes some work and dedication, and it turns out it's a lot harder to *actually* be a paranormal investigator than to merely *want* to be one. What you see on television is either for entertainment purposes only and completely fictionalized or at least edited to make an entertaining program. What occurs in reality is a lot of hard work, and you want to make sure

---

1   A few organizations offer courses in paranormal investigation. Some even offer a certificate upon successful completion. But there's no recognized accreditation process, so literally anyone can hang out a shingle and present such a program. We've been toying with the idea of starting one of our own, and we think ours would be a pretty good one, but even at that, there's no *official* way to distinguish what we'd consider a reputable course (like the one we've been working on) from some random fly-by-night operation.

you're working with someone who's willing to put in the often boring and frustrating hours to see things through.

4) Ask about prior experiences. That doesn't mean they need to have published books like ours or that you need to listen while they describe everything they've ever done. But you're looking for some key pieces of information. First, and perhaps most importantly, have they actually done this before? And if they have, were their results always the same? If your paranormal investigator has never failed to find evidence of something supernatural, they're not doing their job. We're not going to say there's no such thing as a haunted house, but we will say without fear of contradiction that not every house is haunted. Therefore, a legitimate investigator must necessarily, at least sometimes, fail to find the ghost (or demon or alien or whatever else they're looking for). Conversely, if someone has been looking into paranormal claims for a while and has never found something they *couldn't* explain, there's similarly a very good chance that they're just as biased in the opposite direction. Our experience has been that in about a third of our cases, we're able to find a natural explanation for the phenomenon, in about a third we don't witness the claimed phenomenon one way or the other, and in the final third, we witness at least one thing that we can't (yet) explain. Don't necessarily expect another team to have the same ratio in their case files, but they ought to have at least *some* cases in each of these categories.

5) Pay attention to their affiliations. There's not necessarily a right or wrong set of institutional or religious affiliations for a paranormal investigator, but you should be aware of what they are. Our own approach is to remain entirely independent but to maintain good relationships with variety of other professionals and institutions. That way we can access those resources when necessary without having to be limited by their philosophies. Your own mileage may vary. But what you don't want is someone whose particular institutional or religious affiliation is going to color the outcome of the investigation, especially if their affiliations differ from yours. This is most relevant, we think, when it comes to religion. Make sure whoever you work with isn't just trying to push you into their own religious beliefs.

6) Be cautious of teams who work with psychics. We're not going to go out on a limb and say there's no such thing as a psychic. And we're certainly not going to say you shouldn't work with an otherwise legitimate organization simply because one of their members believes him- or herself to be a psychic. However, we caution against taking any information provided by self-professed psychics too seriously (that's equivalent to using one unproven claim to prove another). In our experience, groups that work with psychics, mediums, or other paranormal kinds of people as a matter of course also tend to be strongly biased in favor of finding a supernatural explanation in most or all cases. It needn't be a deal-breaker, but tread carefully.

7) Look for a team with diverse professional backgrounds. This is sort of the antithesis of the last point, because "diversity" of professional backgrounds,

we have to admit, could hypothetically include a claimed psychic. However, what we're really looking for here is a team that can pull from a wide range of expertise. At Rocky Mountain Paranormal, our own members' (past and current) backgrounds have included psychology, photography, law, mathematics, stage magic, history, biology, physics, chemistry, philosophy, religion, folklore, and more. And when we lack the necessary expertise ourselves, our Rolodexes are full of just about every kind of expert you can imagine. The reason this diversity of background is so important is because when we're dealing with anomalous claims, we never know which discipline might provide some key insight, so it's important to be prepared to draw from any or all of them at a moment's notice.

8) Ask about the team's ethical standards. We don't necessarily expect every team to have a written statement of ethics (though it's not a bad idea—see our own essay earlier in this book). But they should at least have thought about it and ought to be able to give you an overview of their ethical philosophy to make sure it matches what you're comfortable with.

9) Beware claims of certification or professional affiliation. As we've discussed, there really are no legitimate certifications for paranormal investigation. Our members are certified in a variety of fields upon which we draw in our investigations, but there's no such thing as a certified paranormal investigator (or if there is, the certification is not from a real accrediting body). Similarly, though there are some confederations of paranormal investigation teams that aim to offer some degree of peer oversight or ethical standards, those organizations are likewise not all created equally and, like the member organizations they claim to represent, tend not to last very long. Possession of a certification or membership in such an collective of organizations shouldn't be taken as an argument against a group, but neither should it be taken as an argument in favor of the group.

10) Insist upon a meeting or interview before the investigation takes place. If the team isn't local to you and travel might be involved, this needn't be done in person, but I would never trust a group that wants to dig into a full-fledged investigation without at least meeting and discussing the matter first. For the client, this is an opportunity to make sure they're a good fit (and to ask some of the questions we've suggested above). And for the investigators, it's necessary to get a better sense of what the paranormal claims are in more detail and to ask questions of the prospective client that might have some bearing on how the investigation should be conducted. Exceptions can be made under emergency circumstances or if the paranormal phenomenon is occurring in real time and you're trying to get someone out to witness it before it stops, but in our experience those situations rarely occur.

If you ever find yourself in need of a paranormal investigation, we hope these guidelines will serve you well, though we could easily come up with another ten or more ideas if we wanted to make this book even longer. Should things not go the way you expect, you should also not be afraid to bring in another team for a second opinion (but at the same time, don't just keep searching until you get the answer you were hoping for, of course). And don't be afraid to reach out to an organization you trust, even if they're not in your region. For example, even if we're not able to actually visit your home and conduct an investigation, we're always happy to consult long-distance and help people in any way we can. Other reputable organizations ought to do the same.

15
The Hall Ghost

Our first private residence investigation in this volume concerned a family of three living in southern Colorado. After living in their home without incident for a number of years, their young child suddenly began interacting with an unseen entity. What began as a seemingly innocent interaction with an imaginary friend soon developed into a terrifying series of incidents involving the entire family.

Figure 15.1. Artist's rendition of the child playing with the ghost. Image: Spraycasso.

There's something about children that seems to go along with tales of the paranormal. In horror fiction, children are often centrally featured characters. Our inner literary and film critics tell us this is probably in large part because horrifying situations that involve the most innocent of people are rendered all the more terrifying. Children are vulnerable and tend not to deserve whatever terrifying thing is about to befall them in the most frightening of ghost stories. But that's fiction. Why children should often be involved in the real-world cases of paranormal claims is a more complicated question.

Perhaps, as some people will argue, children are simply more receptive to the paranormal. Similar claims are often made of animals—they can see things that adult humans seem not to notice. An explanation we've heard for this phenomenon (from the paranormal believers' perspective) has to do either with entities being attracted specifically to the kind of naïve innocence of children and animals or with the idea that adults tend to block out perceptions that might challenge their preconceived notions of how the world is supposed to work.

From a more skeptical perspective, it's also been argued that children are simply more gullible than adults[1] and therefore can be persuaded to believe more readily in things that aren't really happening. Animals, on the other hand, might be seen to more readily react to claimed paranormal phenomena simply because animal cognition and behavior are not identical to humans' and therefore their behaviors can sometimes seem erratic simply because we don't understand what they're doing even if it has a perfectly mundane reason.

Regardless of your philosophy, it is undoubtedly true that the paranormal claims involving children are often some of the most interesting. Of course, they're also the ones about which we always want to exercise the greatest caution.

In this particular case, though the paranormal claims progressed from seemingly innocent to seemingly menacing, we're pleased to be able to report that nothing particularly tragic occurred.

The case involves a family—a father, mother, and boy of about two (perhaps nearing three by the time of our investigation) years of age who we'll call Abe, Bella, and Christian, respectively[2]—who moved into a new home in southern Colorado several years before our investigation and lived in the house that whole time without incident. But after almost ten years, shortly after Christian's second birthday, strange occurrences began for reasons unknown and escalated rapidly. The layout of the house was such that Christian's bedroom shared a common hallway with the living room. Down the hall was the master bedroom, making this central hallway a focal point not only for the family's interactions but the various paranormal phenomena they started to experience.

Christian was the first to experience anything out of the ordinary. One day shortly after his birthday, he was in his bedroom and his parents, spending some time in the nearby living room, heard the sounds of their child playing and mov-

---

1   Perhaps true but we're honestly not so certain.

2   Not their real names. In this as with all of our private cases, the clients' names have been changed and any identifying information has been omitted or altered to protect the clients' anonymity.

ing his toys about. When they started more actively watching from across the hall, they realized his behaviors were not merely those of a child engaged in solitary play, but what they described as a real conversation. Probably thinking the boy was engaged in imaginative play, Abe smiled at his wife.

"Hey," he shouted across the hall. "Who are you playing with?"

Christian paused a moment before replying: "A little girl."

Curious, Abe and Bella made their way across the hall to see if, indeed, another child were somehow in the room. Of course they found Christian seated in the middle of the room, quite alone, surrounded by his various playthings.

Such an incident is not uncommon in the lives of children. They often play at make believe, and had that been the only occurrence, Abe and Bella would have thought nothing of it. Likely, they would have forgotten it completely within a week or two. But a few days later, another happening took place.

In the middle of the night, Abe and Bella were awakened by the sounds of Christian's toy telephone from the lad's bedroom. Annoyed that he was awake and making noise in the middle of the night, one or both of them wandered down the hall to put him back to bed. But to their surprise, the lights in Christian's bedroom were out. When they turned the lights on, they found the room empty. Not just that Christian was in bed rather than playing with his toys. No, he wasn't even in the room where they'd just thought they heard him playing with his telephone. A quick look across the hall and they found the boy fast asleep on the living room sofa.

From that point forward, Christian refused to sleep in his own bedroom.

At this point in life, at just a bit over two years of age, Christian was at the point in his development during which children's linguistic skills begin to increase rapidly. Typical children at age two are capable of stringing a couple of words together; by age three, they're able to form simple sentences. However, they're not necessarily able to express their most complicated emotions or experiences. Therefore, when Abe and Bella asked Christian why he had taken to relocating himself across the hall to the living room couch instead of his own bed every evening, his reply was both simple and chilling:

"Because of the monsters."

Unfortunately, further elucidation was not forthcoming. Likely still uncertain at this point exactly what was happening—did Christian actually see something they should worry about or was this merely a childish flight of fancy?—Abe and Bella started paying much closer attention to everything happening in and around their house. It wasn't long before they started experiencing some paranormal events of their own.

Several friends of the family began to report seeing "strange shadows" moving just at the very edge of their peripheral vision. Some of them said they felt "uncomfortable" in the boy's bedroom. Neighborhood children invited for sleepovers would also refuse to sleep in Christian's room.

Within the family, Bella was next to experience something she thought might be paranormal in nature. It first manifested as nightmares. In her dreams, strange shadows floated and flitted about her bed. More than once, she awoke—and awoke her husband—screaming.

"They're at the end of the bed," she would say.

Abe would dutifully look around the bed but never saw anything. Clearly, Bella was genuinely frightened, though. When Abe returned his attention to her, her eyes were wide open and she was shaking in fear. She explained that she saw two distinct shadowy figures and that while neither of them made any overt threatening moves, she was terrified of them. And while this didn't become a nightly experience, it happened on multiple occasions.

Yet somehow Abe seemed immune to the paranormal goings-on for a much longer time than the rest of the family. Christian wouldn't sleep in his bedroom. Bella experienced nightmares that seemed to bleed over into the waking world. But Abe didn't experience anything for some time. Until one night while he was sleeping. Per his usual routine, he was asleep on his side facing the hallway into Christian's room. On this particular evening, for reasons he's not clear about, he woke up in the middle of the night and "just sensed" something.

When he opened his eyes he saw, just a few inches from his face, a little girl standing at his bedside. After a moment, she simply vanished into thin air.

It was at this point the family decided to seek help and reached out to Rocky Mountain Paranormal. It's not hard to imagine why. The story they told us (and which we've summarized above) really does sound like the opening act of any number of horror movies. Were this a work of fiction, we'd expect the spirits to become increasingly threatening over the next few nights. The good news, of which we quickly assured the frightened family even before beginning the investigation itself, is that unlike in horror stories, in the real world there has never been a documented case of a paranormal entity causing direct harm to anyone. Legends and lore might hint at such things, but we've never come across any such case; in our experience, the harm done—if any—is caused by the people's reactions to what they experience rather than by the experiences themselves.

Nevertheless, we arranged an investigation. We interviewed the family to get all the details of the story. Mostly we did this with Christian not present, though we did ask him a few questions (under his parents' supervision, of course) to get his side of the story.

But after setting up all of our equipment and monitoring the house, we never experienced anything. Neither did any of the family members while we were present. That makes this one of those sad cases in which we struggle to either confirm or debunk anything because it simply didn't manifest itself to us, so our approach was simply to again reassure the family that they shouldn't have anything to worry about and ask them to keep us informed if the events kept occurring. They didn't call us back, so we have to assume at some point things settled down.

As for our conclusions? This is one of those cases in which we have none. The personal experiences of the witnesses are subjective and impossible to measure. We found the family to be credible and levelheaded people, but we can only work with what happens in our presence.

All of the occurrences described could be explained, hypothetically, as psychological phenomena rather than supernatural ones. The explanation could go something like this. Christian, being a young child, engaged in a flight of childhood fancy and conjured up an imaginary friend. When his parents reacted to his behavior, that could have prompted his fear of his bedroom. His own fears and those of his parents could have precipitated the uneasy feelings reported by others. And the stress of the entire situation could have induced nightmares—perhaps even night terrors[3]—in the parents. But we're not saying that's what *did* happen. It's just one possible naturalistic explanation.

At the end of the day, Abe, Bella, and Christian had the kind of personal experience most of our members have been chasing for decades. Whether it was paranormal or not, we imagine it's the kind of story they can tell at parties for years to come.

---

3   Night terrors are a psychophysiological phenomenon in which the brain partially awakens while the body remains paralyzed (as normally occurs during sleep so we don't act out our dreams). The mind is conscious but not yet fully awake and this leaves the individual aware of his or her surroundings, unable to move, and often vividly hallucinating some kind of monster (usually one of cultural relevance to the individual).

16
Enough is Enough

This is the case of a family so terrified of their own home that probably no one could have convinced them to stay. Like the previous case in Chapter 15, it involves a young child who seemed to perceive some things the rest of the family couldn't, though terrifying occurrences eventually plagued the entire family.

Though we were the first investigative team on scene, we were given only one night to conduct our work because the family had already decided to flee the house the following morning.

Figure 16.1. Artist's rendition of the ghost in the attic. Image: Spraycasso.

A point of pride for us in our private investigations is that we're often able to help people get their lives back in order. Whether one believes an allegation of paranormal phenomena to be genuinely supernatural or merely some mistaken natural phenomenon, these kinds of disturbances have a way of often disrupting people's daily lives. Homes, meant to be families' safe and comfortable refuge against the rest of the world, begin to frighten people, and we consider it our honor and duty to do whatever we can to put things right.

In some cases, particularly when disagreement about the paranormal claim has driven a wedge between family members, that means referring people to professional counselors of some sort (whether those might be psychologists or members of clergy or some other related professional). We do this not because we think people are crazy or mentally ill, but simply because the added stress is causing some emotional problem for the individuals involved. Over the years, we've developed an extensive network of professionals who we know understand how to work with paranormal claimants. After all, unquestioning belief in the paranormal could reinforce a delusion if the claim turns out to be delusional; on the other hand, unquestioning denial of the paranormal is not healthy either when applied to perfectly sane people who are simply experiencing something they don't understand.[1]

Often, though, the best way we can help people get their lives back in order is simply to help them understand what's happening in their home. Our case of "The Toilet Ghost" (Volume 1, Chapter 23) is a good example of this. The family were terrified of all the paranormal things happening in their home but we were able to meticulously explain all of the occurrences through a variety of completely mundane phenomena. Not only were we pleased to have been able to solve such a multifaceted mystery but the clients were thrilled to discover they didn't actually need to be afraid of their own home anymore. Even when we can't find a natural explanation, sometimes the results of our investigations—and perhaps our general counsel based on decades of experience looking into such claims—are enough to put people's minds at ease.

This is one of those cases where our primary goal—even above and beyond our longstanding search for evidence of the paranormal—was to try to help the family make peace with their home. Sadly things didn't work out so well in the end.

For us, this case began with a panicked call from a man we'll call Dale who,

---

1   It's actually distressing to us how many so-called "professionals" adopt one of these two extreme views rather than treating their patients and their claims with the respect (but not utter deference) they deserve. Training for psychologists, clergy, first responders, and other professionals is understandably limited when it comes to topics of the paranormal (or, indeed, different religious beliefs and practices those professionals may encounter in the field). But though we find that limited training completely understandable, the large number of paranormal claims that are made every year also leads us to think that such lack of training is unacceptable. We're always thrilled when we find professionals who take the correct philosophically neutral approach because those are the ones who are most likely to actually help people.

as our introduction to this chapter would suggest, was absolutely terrified of his own house. Before we get to our own involvement in the case, we'll report the backstory as it was relayed to us in our interview the family members.

Dale lived in western Colorado with his wife Eugenia and three-year-old daughter Franny. They'd just moved into a new rental property. It was one of those too good to be true kinds of deals we've all read about in horror novels. Their prospective home was far cheaper than other rental properties of comparable size and location. Furthermore, it was vacant and ready for immediate move-in. A dream come true. Seeing it as a rare opportunity, Dale immediately began the rental process.

The rental company's manager handed him the keys and told him to walk through the property with his family and return the keys afterward. In hindsight, Dale thought it strange that they'd been given the keys and told to visit without supervision, but we can understand how someone pleased with his good fortune might look past little oddities. After the walkthrough the family were thrilled. When they returned to the rental company's office they signed the papers, paid the first month's rent and security deposit, and began to move in.

Things were great for a few days, but it wasn't long before their dream home turned into a nightmare.

The first sign the house wasn't quite perfect was some clean-up they had to do before they could really move in. Prior tenants had left a variety of belongings behind. One of the back bedroom closets contained a stash of adult diapers and clothing, unused syringes, and other assorted medical devices. Nothing particularly disgusting or troubling, but also nothing they wanted lying around with a child in the house. Other bedrooms contained children's toys and clothing. It took the family about a day to clear out the old items and about a week to finish moving their own things in.

One evening shortly after they'd finished moving, the family were sitting in the living room. The front door was open to let the cool night air replace the day's heat. Franny wandered over to the screen door and parked herself there, staring out into the thick bushes surrounding the front stoop. Eyes squinted, she tilted her head to the side.

"See Dale," Franny said. "See George."[2]

Dale rose from his chair and went to the open door expecting to see someone, perhaps a new neighbor, out in the yard or approaching the front door. No one was there. He quickly scanned the entire yard before focusing his attention on the point Franny was pointing out.

"See George," the girl repeated, staring up at her father.

Dale alternately squinted and widened his eyes to adjust his gaze but never did see anyone out in the yard. Perplexed by her father's incompetence to see what was clearly so obvious to her, Franny took matters into her own hands. She

---

2   George was also not the name the child used. We don't know who she was actually talking about and whether the name she used was someone she knew or not. We've chosen to fictionalize this name as well, even though it's not a family member, on the off chance the name could have belonged to a real person somehow involved in the case or with the property.

returned her attention to George, planted her tongue firmly between her lips, and began to blow raspberries at the unseen visitor.

Over the next few weeks, Eugenia began to notice that Franny was becoming increasingly chatty when playing alone in her room. On one such occasion, Eugenia asked Franny with whom she'd been playing. The girl looked at the window and said, "the boy." She then put on her backpack, turned again to the window, and said "come on, let's go" before leaving the room. As was the case in our previous chapter, the parents were somewhat troubled by this new behavior, but their first assumption was that Franny had simply developed an imaginary friend as children often do.

Then one night, Franny's behavior took a turn to the more disturbing. In the middle of the night, she awoke screaming and bounded out of bed, running past a confused Eugenia as if she didn't even see or recognize her own mother.

After some time, Dale and Eugenia finally managed to calm the girl down. But because of her young age, they were unable to get her to articulate what had frightened her so.

Soon after, the parents began to experience strange things for themselves. It began with a sound they described as "chains rattling" in the attic. They looked but no one was there, and they couldn't find any chains. A few days later, they heard what sounded like someone rummaging about in the kitchen followed by footsteps coming down the hall. Again, no one else was present and nothing in the kitchen seemed to be out of place. On another evening while lying in bed, they heard music that sounded as if it were being played on an antique record player. Accompanying the music was an indistinct voice whose words they couldn't quite make out.

A search began. Dale went through the entire house, room by room, to figure out who or what might be making the sounds. He didn't find anyone, but did notice that the bathroom door was moving back and forth, ever so slightly, apparently on its own. As quickly as it began, the music stopped and Dale never figured out its source.

Shortly thereafter, Dale and Eugenia were in the living room when they heard a screeching sound. At first they thought it was a cat, but as they listened more attentively they came to the conclusion that it sounded more human. Dale couldn't make out any particular words, but Eugenia believes she heard the screaming voice say "help me, mommy; help me!"

Of course they investigated, not only beginning to suspect something paranormal but also concerned that perhaps a neighbor could be in danger. But they couldn't pinpoint the screaming's source. They believe it sounded like it was coming from inside their house, though they dutifully searched both inside and out. Every time they thought they were getting close to the sound's point of origin, though, it seemed to move to another location.

Eugenia said that one night as she was falling asleep she saw a shadow that resembled a dog. Initially she assumed her eyes were playing tricks on her as she drifted into sleep, but then she later awoke to the sounds of a dog pattering around the side of the bed. This time, though, she didn't see anything. From that point, all three members of the family would occasionally pick up the distinct aroma of a rotten dog smell throughout the house. They didn't own a dog.

Phenomena seemed to increase in intensity and frequency over time. While Eugenia rarely saw anything disturbing for the first few weeks in the house, she increasingly began to notice shadows lurking just in the corner of her eye. While in the back room one night she turned and saw—directly rather than peripherally this time—a pure black, tall and thin shadow figure standing in the room. She said it was clearly humanoid but pure black with no distinct features. She ran from the room, headed to the master bedroom and hid under the covers.

Another evening, Eugenia and Franny were watching television together in the living room. Dale emerged from the bedroom and headed down the hall to join them. As he passed the bedroom door, he noticed it was partially open and saw Eugenia beginning to rise from the toilet through the open door. To be polite, he passed quickly. However, both his wife and daughter were still in the living room. He ran back to the bathroom, threw open the door, and found absolutely nothing and no one.

This was the point at which they said "enough is enough," and planned to leave the house. They called us in a panic unsure of who else to call or what else to do. They told us they planned to abandon the house as soon as they could.

Situations like this are about as close to an emergency as we encounter in paranormal investigation. From our initial phone call with Dale we knew they were planning to abandon the house. We also knew that walking away from the rental—combining the cost of staying in a motel until they could find another property with the loss of their security deposit because of the unannounced departure—would devastate the family's finances. We made plans to investigate the property as soon as possible and encouraged them to hold off and give us a chance to look into the matter before they made any rash decisions. However, they said we had only one night because they were moving out the next morning.

Our small team (consisting of as many members as could drop everything for an unscheduled emergency investigation) arrived in the evening and conducted a more detailed interview, from which we drew the events described above. We then spent the night at the house in the hopes of witnessing or recording some or all of the various events they'd been experiencing. Nothing materialized that night.

We tried to encourage the family to give us a few more nights, not only because we wanted to complete our own investigation but, more importantly, because we hoped our eventual findings might be able to save them from the financial consequences of abandoning the house. But they wouldn't budge on the subject. Our weary team left the following morning with the terrified family—all their worldly possessions hastily packed into the family car—right behind us.

After any investigation, successful or not, we discuss and try to figure out what we can learn from the experience. This ended up being one of our most frustrating cases because not only were we unable to discover the source of the paranormal claims, not only were we unable to save the family from abandoning their home, but we were also unable to come up with any kind of philosophical or methodological lesson about what we might have done better.

Honestly, the only thing we can even think of that we might have done in addition to what we actually did was to buy or rent the house for ourselves, but we lack the financial means to make those kinds of commitments. That said, if you think your house is haunted and you're about to walk away from a lease or a mortgage, at least give us a call first and try to give us time to figure something out. If we can't save you from financial ruin, we'd at least like the opportunity to call our own real estate people and see what kind of bargain we can get on the haunted house of our dreams.

17

The Home of Death

Sometimes a property's history is even more frightening than the paranormal phenomena themselves. Such was the case in what will be one of our shortest chapters despite being truly the stuff of nightmares. It all centered on a ranch house in a rural area east of Denver that turned out to be the site of more than a single home's share of death over the years.

Figure 17.1. Artist's rendition of the Home of Death. Image: Spraycasso.

This will, unfortunately, be a brief one. Yet despite not really having a whole lot we can report about this case, it remains one of the ones that's stuck in our memory.

What we came to call the Home of Death was purchased shortly before our investigation by a family consisting of father Harold, mother Irene, and two teenaged sons Jack and Kevin. The home was attached to an active ranch and so the family shared their property with a variety of livestock including cows, sheep, and their small herd of pigs.

Rumors about the home's history were known to the family but unconfirmed. They'd heard that someone had been shot on the property, but didn't necessarily think anything of it until strange things began happening.

Throughout their time at the house, they often heard people wandering the rooms and halls at all hours of the day and night, even when no one else was present.

A particular spot on the ceiling, they said, had to be constantly painted and re-painted because no matter how many times they covered it up, a mark resembling a bloodstain would soon reappear. They believed that to be the location of the rumored death in the property.

The most spectacular occurrence took place when they purchased a new pig house for their swine. This consisted of a feeding trough and a covering so the pigs could enjoy the shade or take shelter during storms. After the family purchased it, they directed the mover to place it inside the pig enclosure, which he did. The following morning, the entire pig house was no longer in the pig enclosure. It has been moved to the other side of the fencing near the home. Though not as large as an actual house, the pig house was a large enough structure that an overnight operation to move it outside the fence (and without disturbing the rest of the property) would have been quite the undertaking, made all the more improbable by the fact that no member of the family heard any disturbance during the night.

The thing that really troubled the family, though, and what prompted them to seek our help, was that since they moved into the house, they started fighting far more than was typical. Whenever they were in the house—and *only* when they were in the house—they were constantly at each other's throats (metaphorically speaking, we hasten to clarify). Combined with the unusual happenings they'd witnessed and the home's reputation for being the site of a death in the past, this made them think they were perhaps being influenced by some kind of malevolent force and they wanted our help in figuring out what might actually be going on.

We arrived for our investigation in the afternoon and they showed us around the property. During the tour, they pointed out the various sites of interest. In particular, we took notice of the spot on the ceiling where they said a bloodstain regularly reappeared. But at the time of our investigation, no stain was present. Neither did one reappear during our night on site.

They also pointed out the pig house, now back in its proper place behind the fence in the pig enclosure. Of course we weren't present when it actually moved so our investigation was necessarily limited in that regard. But we did note that the pig house itself was a round structure about eighteen feet in diameter and six to seven feet tall. That pretty much ruled out the hypothesis we were considering that perhaps some local teenagers or other pranksters thought it would be funny to move it. Not that it's impossible, but it would have taken a large crew, and in our experience relatively few pranksters are organized enough or willing to go to that much trouble.

After our tour, we set up our usual array of monitoring equipment and spent the rest of the afternoon and all night at the house. Nothing unusual happened while we were there. We did spend some time inspecting the structure around the part of the ceiling where the blood stain was said to reappear, hoping to find something like a slow water leak that might explain the phenomenon. After all, rust stains from a leaky pipe could easily be confused for a blood stain.[1] There didn't seem to be any evident culprit in the area. While that doesn't mean we can rule out this kind of natural explanation, we did at least look and didn't find anything.

After our on-site investigation, during which Harold and Irene told us about the home's reputation for being the site of a death before they purchased it, we did some diving through the public records surrounding the property to see if we could either confirm or debunk those rumors. And we hit the motherlode.

It turns out, this house wasn't the site of just a single death. Oh, no. It was the site of more violent deaths than any private residence we've ever encountered, and more even than most of the public ones if we exclude places like hospitals (see our next volume) or jails (see Volume 1, Chapter 10). We were able to locate records confirming the house was the site of *at least* two suicides, one murder/suicide, and one "suicide by cop." This is not typical and after we made that discovery we started calling it the Home of Death.

That leaves us with a bit of an interesting conundrum, particularly since we weren't able to observe and either confirm or debunk any of the paranormal allegations around the house. It's absolutely true that the house is home to more than its share of violent death. That in and of itself does not mean anything paranormal is happening there, of course, but we're also aware that believers in the supernatural could easily interpret this as being the result of some kind of malevolent force or entity causing the home's residents to behave violently. That our own clients expressed that they fought uncharacteristically when in the house but were fine elsewhere certainly leans into that kind of belief. But it's just not enough evidence for us to say that's what's happening. It could be that some

---

1   Though initially red, stains from dried blood quickly turn a slightly orange brown color. This is because the red blood cells contain iron-carrying hemoglobin. When iron is oxidized, it forms iron oxide, also known as rust. That is, rusty metal and old blood have undergone a similar chemical process. Therefore it can be quite difficult to tell them apart at a glance. Over time, rust tends to retain more of its reddish-orange coloring, while blood stains become a darker brown (sometimes almost black) as the hemoglobin breaks down into a compound called methemoglobin. Actually distinguishing blood from rust, though, is not always a simple matter and should be referred either to a forensic expert or conducted through chemical analysis. Substances such as luminol (see Chapter 5) can be used in such analysis.

environmental toxin could be causing disturbed behaviors.[2] It could be that the rumors about the house themselves cause people to be "on edge" which could alter their behaviors. It could even be pure coincidence.[3]

When we explained our findings to the clients, we emphasized the importance that they should seek psychological counseling to deal with the fighting and emotional troubles they were having. Our argument was that whether the cause was natural or supernatural, at least they could get professional help to treat the symptoms. They agreed. We also told them to report back to us if they experienced any further strange phenomena or needed any further help. They never reached back out, so we can only hope and assume the counseling was successful.

---

[2]   We didn't find anything during our investigation to suggest it, but we have significantly expanded our environmental monitoring capabilities in the years since this investigation.

[3]   Coincidence is never the most satisfying explanation but we have to always be aware that coincidences really do happen. And given the number of people in the world, unlikely things will happen to people purely by chance all the time. In probability theory, this is known as the law of truly large numbers (attributed to mathematicians Persi Diaconis and Frederick Mosteller). A more intuitive way to think about it is to realize that, since there are (as of this writing) approximately eight billion people in the world, one-in-a-million coincidences should be expected to occur to eight thousand people at any given time.

18
The Invisible Jellyfish

This might be one of our strangest ones. We always like to say that we love when a case comes along that isn't focused on ghosts or aliens. Not because we don't like those—of course we do—but simply because it's nice to get one that's different from the claims we hear in the bulk of our cases. This particular case involves—we kid you not—a house infested with invisible jellyfish monsters. When people ask about the diversity of claims we investigate, it's one of our favorite examples of something completely unexpected.

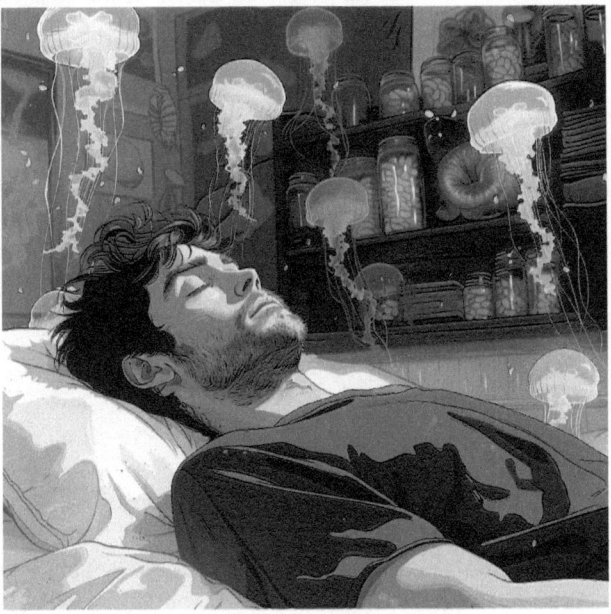

Figure 18.1. Artist's rendition of the Invisible Jellyfish. Image: Spraycasso.

Among paranormal investigators, we're something of an anomaly. Most teams, even if they have a general interest in all things paranormal, quickly find that they specialize in a particular area of the field. Some hunt for ghosts. Others track UFOs. Still others camp out in the woods looking for Bigfoot. Few do what we do and generalize. We're happy to look into *any* kind of paranormal claim. Sometimes we're asked about the most unusual claims we've ever investigated, and the one in this case is certainly a contender.

We were contacted by a man we'll call Lou who was in his mid-50s and lived alone in a very small house. He told us his house was infested by "entities" and he wanted our help in investigating to see if we could determine what was causing the events he'd been experiencing. During our initial correspondence, his story was pretty short on details, but he seemed troubled enough by the phenomena that we thought it best to arrange an on-site meeting to get more information in person and begin looking into the case.

When we arrived, we found a house that wasn't exactly unclean but we did notice that beer cans mason jars half-filled with water were scattered throughout the entire house. That certainly caught our attention. When we asked about the jars, Lou said they were related to his paranormal phenomenon and that he'd explain the whole thing so we sat down for an interview.

The house, Lou explained, was infested by invisible jellyfish-like creatures. We have to admit, we weren't expecting that one! From our initial correspondence, we were prepared for a more run of the mill ghost story. This was something different. Lou went on to say that he didn't know exactly what the creatures were except that they were invisible but seemed to have similar structure to jellyfish. They would float throughout his house all the time, but made their presence known primarily at night. When Lou slept, they would hover over his bed and use their "tentacle stingers" to "shock" him on the legs.

Though it seems unnecessary given that these creatures aren't alleged to be actual jellyfish but instead something that merely resembles jellyfish, we'd be remiss if we didn't include a brief biological note. Jellyfish (or more accurately sea jellies) consist of umbrella or mushroom like "heads" or bodies trailed by a number of tentacles. The tentacles are known for their ability to sting. While they do not actually possess the power to deliver electric shocks, their stings' symptoms could plausibly compare to an electric shock. Most often, these symptoms include severe, instant, and persistent burning or stinging pain combined with inflammation. This is caused by stingers along the length of the tentacles which each contain a microscopic bulb of venom that's injected when the victim touches the tentacle. Whole-body effects, including (rarely) fatal ones, include dizziness or heart failure as a result of the sting. Most stings are similar in severity to a string of bee stings and will self-resolve over the course of several days during which home remedies may alleviate some of the symptoms.

Carrying on with Lou's story, he went on to explain that this had been happening for several months and he'd begun to experiment on his own. The mason jars filling his home were there because he'd been able to capture some of the smaller creatures and keep them in the jars.

One jar he showed us was half filled with water and had a pencil in it. He explained it contained a dead jellyfish creature. He knew this because when he stirred the water with the pencil, he could feel the resistance as the writing implement brushed against the creature. We tried it—why not?—but didn't feel anything.

Then he showed us some of his video footage. Over time he'd collected hundreds of hours of video of his experiments on his invisible jellyfish creatures. The particular video he played for us showed one of the mason jars in close-up, again a bit more than half-filled with water. On the film, he placed a small worm into the jar which then floated through the water. At one point the worm momentarily disappeared. We knew this was due to the curvature of the glass and the refractive properties of light simply making the worm disappear for a moment, but Lou (still on the video) got very excited.

"See, he just ate the worm," he exclaimed.

A second or two later, the worm reappeared as it moved past the refractive "blind spot" in the jar.

"He didn't like that one," Lou said. "Look, he spit it out."

Even without conducting our own experiments on the jars, we knew exactly what we were seeing. Exploring the refraction of light is a classical science fair project and is a well-known principle of optics. Still, we watched a few more of Lou's videos as he continued to explain what he believed he'd been experiencing.

Our entire interview with Lou lasted just about thirty minutes. In that time, we noted that he'd consumed seven beers and five cigarettes. Clearly he was stressed out about what he thought he was experiencing. Equally clearly, his routine of self-medication with alcohol and nicotine wasn't improving matters.

This is a delicate situation for us because it was clear that Lou was struggling with mental issues and that this, rather than invisible jellyfish-like entities floating throughout his home, was at the heart of his real problem. But saying so isn't necessarily the most helpful move.

Something we've often repeated is that not everyone who believes in the paranormal or has a paranormal experience is crazy. But when you work in the paranormal for more than twenty-five years, you're going to encounter some people who, while "crazy" is not a polite or medically precise term, are struggling with a variety of mental problems. The solution for those people is to get them into the kind of professional care they need, but if one were simply to say "there's no such thing as an invisible jellyfish and you should see a psychologist," they're likely to clam up, close off, and shut down the conversation. Even in cases where the paranormal claim seems more plausible, we often refer people to psychologists under the theory that however the paranormal investigation might turn out, a professional ought to be able to help them deal with the stress of the situation. That's completely true, and we mean every word of it—even if your house really is haunted, or you really were abducted by aliens, or whatever your paranormal claim is, we still think seeing a psychologist is often a good choice for a large proportion of our clients.

In Lou's case, we decided to take the same approach. Though we didn't believe his paranormal claim and had come to the conclusion that his experiences were induced by stress and substance abuse, we didn't deny it outright. Instead we noted his clear stress about the whole situation and told him we'd be glad to help in any way we could, but that the first step would be for him to see a psychologist to work on the stress so that he'd be better prepared to work with us.

Lou agreed with that idea and said he would seek help and then call us back if there were any way we could help after he got his stress more under control. We never heard from him again. We hope he was able to get the psychological help he needed.

At the end of the day, we're glad to have a truly unusual claim to add to our casebooks even though it turned out to be a mental issue rather than a genuine infestation of invisible jellyfish. Most importantly, this case illustrates the diversity of interactions we deal with. While most of our clients are completely rational and levelheaded people who simply don't understand what's happening, we do occasionally find ourselves working with more disturbed or mentally ill people. The key lesson in such circumstances is to always treat people with respect, whether you believe their paranormal claims or not, and then work to either conduct your investigation or to get them into whatever kind of professional care they need without either reinforcing a belief you don't think is true or doing injury to the client's dignity by denying the experience out of hand.

Interestingly enough, "invisible jellyfish" is only one of two standard answers we give when asked about the most unusual or unexpected claims we've heard. The other is "giant cricket demons." The solution to that case was quite different from this one, and we're planning to include it in Volume 3, so stay tuned.

19

Had to Leave the Ghosts Behind

Several of the private investigations in this volume follow some common themes. We mentioned earlier why we often prefer to be the first team to investigate a property, and this case provides a good example of how another team may have caused some damage. It also follows the theme established in Chapter 16 of families becoming so terrified of their paranormal experiences they flee their home.

The case involves allegations that the home was haunted not by one but by several distinct spirits who manifested in a variety of ways.

Figure 19.1. Artist's rendition of the ghost in the kitchen window. Image: Spraycasso.

We mentioned before that we like to be the first team to conduct an investigation. When it comes to more public (or open to the public) kinds of venues, this is less important and often impossible in the first place. Sure, it would be nice to be the first team to "break the news" of a particular haunting, but largely those places are pretty well-explored and their ghost stories are pretty well-documented by the time we arrive. But in the case of the private cases, it's often much more possible to be the first team, and we think its more important because, as we've mentioned, some teams who are less careful than we are can do a lot of psychological damage[1] before they leave. This case may not be the most extreme of such examples we've ever come across, but it does illustrate how we sometimes have to clean up another team's mess.

The home in this case was a single-family rental home within the Denver Metropolitan Area (we don't want to specify exactly which city) occupied by a single mother in her mid-thirties we'll call Marjorie and her approximately six-year-old daughter Nancy. During our investigation, Nancy was staying with relatives out of state, but was present for much of the background story that led to paranormal suspicions.

Claims about this house were numerous but followed many of the same kinds of patterns we've seen in a large number of supposedly haunted houses. Footsteps sounded down the main hallway when no one was present. Marjorie and Nancy sometimes saw reflections of people in mirrors and windows, again even though they were alone in the house. Sometimes similar images would appear in webcam footage. Other times, they'd see shadowy figures with the naked eye, lurking in the hallway, main living room, and bathroom. Furniture sometimes shook. Marjorie said she was sometimes touched by an unseen entity. For Nancy's part, she was terrified of the "man in the hallway" and often saw what she claimed were ghosts around the house.

Though clearly the kinds of things one would want to look into, none of these would particularly make this case stand out among the rest. What ended up making this one more interesting was how the story progressed once Marjorie decided to call for "expert" assistance.[2] We were not the first team she called, however.

The team who worked on Marjorie's house prior to us were people known to us. In fact, we've come across their work on several of our own cases. To protect everyone's identity, we'll not identify them by name, but simply call them the "other team." For what it's worth, our experience with them has led us to believe they're well-intentioned and quite dedicated individuals, but also that their methods differ from ours in some important ways that we think call their conclusions into serious question.

This other team responded to Marjorie's call for help and dispatched a team

---

1   Or in some cases, property damage, but that's a whole other kind of issue.

2   As noted earlier, we encourage caution when dealing with anyone who claims true expertise regarding the paranormal. Even after a quarter of a century of work in the field, we don't really consider ourselves experts because the field remains so ill-defined that true expertise is elusive.

of their own investigators who chose to use a mix of more scientific apparatus and self-professed psychics with whom they work. Much of the information
Marjorie would eventually tell us about the case came from this team's psychics,
meaning, of course, that we got second-hand reports of already questionable
information, so take it all with as much salt as may seem appropriate.

According to this psychic report, the house was home to six distinct ghosts:
a "Hall Ghost," an "Old Man," two children spirits in the back yard, a woman,
and a sixth man who had hanged himself in the bathroom. However, they later
changed their opinion about this last ghost and decided that rather than an actual
spirit, he might have been a past life reflection.[3] Obviously, because these reports
came from a psychic, we have to report them as part of the paranormal lore while
also remaining skeptical of their validity.

On the more scientific-looking side of things, the same team also walked
through the house with a "meter" (we assume this was probably an EMF meter,
but Marjorie wasn't certain). Though the team reported this meter indicated the
presence of supernatural entities, we again have to take the evidence with plenty
of salt because a) we're not even 100% sure what kind of meter it was, and b) if it
was an EMF meter, we've known plenty of ghost hunting teams to misuse these
devices and misinterpret their data.

After collecting their initial data, the team sat around the table with Marjorie
to discuss the case for some time. They then claimed they were able to "remove"
two of the ghosts by burning sage around the house and "putting them into
a melting pot." Sage burning has long been a part of paranormal lore and is
thought by many to be useful in protection against evil spirits (we discussed it in
more detail in Volume 1, Chapter 16). Putting ghosts into a melting pot, however,
was new to us. We're still not quite sure where that idea came from or what it
entails. Were we ghosts, though, we're pretty sure we would find it objectionable.
It sounds painful.

Ghosts dutifully secured in the melting pot (whatever that means), the team
then informed Marjorie that she'd have to deal with the other spirits on her own
because they weren't something that needed to be removed. For example, she
was to shout at the two ghostly children in the back yard whenever they got too
active because this would upset them and would be sufficient to get them to leave
her alone.[4] One of the other ghosts was supposed to be "attached" to a cigarette
lighter in the main room. It was not made clear whether removing the lighter
would remove the spirit or just anger it.

---

3   We haven't done a lot to investigate claims of past life regression or reincarnation
simply because it's incredibly difficult to come up with rigorous investigative protocols
for these things. We have read the popular reports on the subject and found some of
them interesting but, without those proper controls, ultimately unconvincing. With
regard to past life regression *therapies*, we're even more skeptical because it's been well
documented that such practices can involve, rather than past life recall, the implantation
of false memories. We recommend the works of Dr. Elizabeth Loftus on the subject.

4   As people who regularly engage in unusual activities in our yards, we can attest that
yelling at invisible children would be the kind of thing that would attract the neighbors'
attention.

Their work finished, the team left Marjorie to yell at the ghosts on her own time. But it didn't work. Instead of solving her problem, she said the activity only increased after the team left. She started seeing the same kinds of activity she'd reported before in the hallway and bathroom, but with increasing regularity. On one occasion, she said something grabbed her. On another, something invisible got into bed with her. Whatever the other team did, it didn't seem to sit well with the spirits.

Enter Rocky Mountain Paranormal. When all of this was happening, we were up in the mountains on another investigation and received a frantic phone call at about 3:00 a.m. Because we were investigating, we happened to be up at that hour and accepted the call. Of course it was Marjorie. She was distraught to the point of terror and told us another team had just left her house after confirming it was haunted, but that they'd left her to deal with it on her own and she didn't know what to do. We talked with her for a while to start to put her mind at ease and scheduled a time to visit the location as soon as we were back in town from our current investigation.

This is the kind of thing we always want to avoid. We don't encourage people to call us first out of ego. It's simply that we know how frightening it can be if a supposed "expert" confirms your house is haunted—or worse, infested with demons, as has also happened to several people with whom we've worked—only to then walk away without lifting a finger to actually explain the phenomenon or solve the problem. True, sometimes investigations fizzle out without real resolution, but a legitimate paranormal investigator should at least see things through as long as possible and refer people to other types of help if nothing else works.

Once we got back down from the mountain, we arrived at the house at the scheduled time. As we got out of the car, we noticed the curtains at the front window (once we later got the layout of the house we were able to confirm it was the kitchen window) were drawn and someone walked past in the house. But when we knocked on the door, no one answered. That was strange enough, but we didn't think too much of it until Marjorie arrived a few minutes later and let us in. There was no one else in the house. To this day, we're not entirely sure who or what we saw, and unfortunately because this occurred just as we were arriving, we didn't yet have any video cameras running.[5]

Before Marjorie arrived and the house's emptiness gave us a Grade-A case of the wiggins, we took the time while we were waiting to gather some EMF readings around the property. We noticed higher than usual readings, but not from any paranormal cause. Rather, the house was directly beneath two major overhead power lines. About eight feet from the home, EMF readings dropped dramatically, but inside and just outside the structure, levels were much higher than normal.

After conducting an interview with Marjorie (during which we obtained most of the details of what the previous team had done) and getting a quick tour

---

5   These days, we'd hope such an event would show up on one of our cars' dashcams if we didn't have any other recorders going. And on some investigations, some of our crew might wear body cams much as police officers do. But this investigation took place several years ago, long before such devices were popular.

of the house, we set up our monitoring equipment. We paid particular attention to the EMF levels throughout the house because of the power lines and elevated readings we'd obtained outside. Our thinking was that EMF levels, even if from a completely explainable source, might be an important piece of the puzzle in light of their effects perception and cognition in a certain proportion of the population. We did find elevated EMF levels in Nancy's bedroom and the hallway, as these were close to power lines and the breaker box. Readings in other rooms were more in line with what we'd expect given the distribution of appliances throughout the house. At one point we shut off all the power in the house to see if we could detect any natural electromagnetic fields that were being drowned out by the manmade readings. We didn't find anything particularly interesting during this experiment, but EMF levels were still higher than what one would normally find in a residence (but that's because of the external power lines, which we of course weren't affected by us turning off the home's breakers).

We do think it's possible (not certain, but possible) that the EMF levels in the home could account for at least some of the paranormal phenomena Marjorie and Nancy claimed to witness. It's unlikely they would explain all of the paranormal claims, and EMF levels of the kind we detected would not be medically dangerous, but they certainly were high enough to cause some hallucinations or mental disturbances in people with the predisposition to be sensitive to electromagnetic radiation. We wonder (without confirmation) if these readings where what the other team took as indication of a ghost on their own meters.

Later in the evening, when Marjorie was preparing to go to bed, we set up infrared cameras so we could continue our video monitoring in the darkened house. The next morning, Marjorie told us she felt like someone had gotten into bed with her, but we checked our video monitoring and found no signs of anything out of the ordinary and she slept through the night except for one brief interlude at about 5:00 a.m. when the landlord arrived to collect rent. Once he left, Marjorie went back to sleep.

Other than elevated EMF levels and what- or whoever we saw in that kitchen window upon our arrival, we didn't witness anything unusual during our night in the house. Our reassurances seemed to make some progress toward calming Marjorie's fears, and we suggested we should schedule additional on-site follow-up investigations to see if we could learn anything more. However, while she wasn't as panicked as she had been when she first called, Marjorie said she had already decided to move her and her daughter to another home. She just didn't feel comfortable there anymore.

Unfortunately, since our access to the house left with Marjorie, we, too, had to leave the ghosts behind.

20

The Ghost in the Baby's Room

Though our write-up here in this book won't be nearly as long, this represents our longest private investigation to date, clocking in at five days of intense investigation. Some of the public investigations, of course, are longer than that (as you'll recall from Chapter 14, some of those have lasted on and off for more than a decade), but private investigations are usually in and out matters. This one feels different not only because of its longevity but because of some strange occurrences that happened during the investigation. It took place at a private home in a Denver-area suburb and remains one of our more intriguing mysteries.

Figure 20.1. Artist's rendition of the ghost in the baby's room. Image: Spraycasso.

In keeping with our theme of being willing to see an investigation through to the end rather than scaring people of their own homes and then hitting the road like *certain other* "investigators" have done, this case represents a full five days of on-site research as we worked to figure out what was happening in this house. By the end, we were still left with some mysteries, though we were at least able to come up with a plan of action that we thought would be useful for our clients.

The action took place at a single family residence in a suburb of Denver. The family consisted of a father and mother who we'll call Oswald and Pauline, respectively, and their late-infant or toddler son who we'll call Quentin, along with a family dog. As far as we know, we were the first team to investigate their house.

When they reached out to us, they listed a variety of claims in no particular order, and we struggled to get a sense of the timeline over which these events developed. However, they were quite clear about what events had been occurring throughout their house. First of all, Oswald and Pauline consistently heard the sounds of people moving throughout the home at all hours of the day and especially at night. The door between the house and the attached garage seemed to always find a way to open itself even when they knew they'd shut it. They felt like there was "something" in the crawlspace beneath the home.

Those things might be disturbing enough, but they had two main concerns. First, since the activity had started, they said they were always fighting, and it kept getting worse as things progressed. Most concerning of all, they said they kept seeing something sitting in the rocking chair at the side of Quentin's crib. The house was arranged such that the parents' bedroom afforded a clear view of the nursery when the doors were open. Pauline said she was constantly running to the baby's room because she saw this "figure" sitting in the chair. She described it as a dark and shadowy figure with human shape and proportions, "just bigger."

We certainly understood why that would be distressing for this young family so we arranged a time to meet at their home for a more detailed interview and to begin our investigation. Upon entering for the tour, the first thing we noticed was that their housekeeping was, to put it as politely as possible, on the "creative" side. It honestly looked like in addition to the fighting they'd told us about, they had also stopped cleaning the house as soon as the paranormal phenomena began.

Lack of cleanliness is far from a universal trait among paranormal claimants. In fact, some of them are among the most fastidious of people we've known. But we have noticed that many of them do struggle to keep a tidy home. It may not be a universal trait but it's a common enough one. Sometimes it may be that the stress of the paranormal occurrences cause people to lose track of their cleaning. Other times, lack of cleanliness can directly lead to environmental factors that might cause the allegedly paranormal phenomenon. And of course in other cases still there might be no direct causal relationship at all. We're not judging people. Lord knows we all have our own unique faults and blind spots. But we do have to pay attention to these kinds of things in case they become relevant during the course of the investigation, so we tend to notice the condition of a property. Even when people have made an effort to clean up because they knew they had guests coming, we've gotten pretty good at spotting the little signs that give away their housekeeping habits. Our advice if you ever need to bring in an investigator would be to leave the house as it normally is when the paranormal phenomena occur. Perhaps the mess is embarrassing, but as we said, we're not here to judge. We are, however, here to determine possible causes for the paranormal claim, and whether those causes turn out to be supernatural or mundane, anything that's done to change the conditions under which we'll be observing just makes our job harder.

That wasn't the case here, though. This particular family seemed oblivious to the mess they'd created and had been living in. But we didn't bring the subject up immediately. We just silently noted it and moved on with our interview and tour of the house.

The investigation then began according to our usual protocol of placing monitoring and recording equipment throughout the house, with particularly heavy surveillance of the baby's room and the crawlspace since those seemed to be the "hot spots" for paranormal claims in this residence. Once that was done, we asked the family to make sure their dog stayed in the back yard for the duration of the investigation. The yard was fenced and the weather was hospitable enough, so we felt this was an appropriate precaution to keep the animal from accidentally wandering through the house and contaminating the data. It was also convenient enough for the dog because Oswald and Pauline wanted to spend part of their evening outside enjoying a smoke.

We joined them outside while we let the cameras run unattended for a while and asked some additional questions about their background and experiences. In our absence, something weird was happening in the crawlspace.

Once we got back inside to look at the monitors, we immediately noticed something was off. Video from the crawlspace camera showed a completely different scene from the way we'd set it up. Yes, it still showed the crawlspace, but from an entirely different angle.

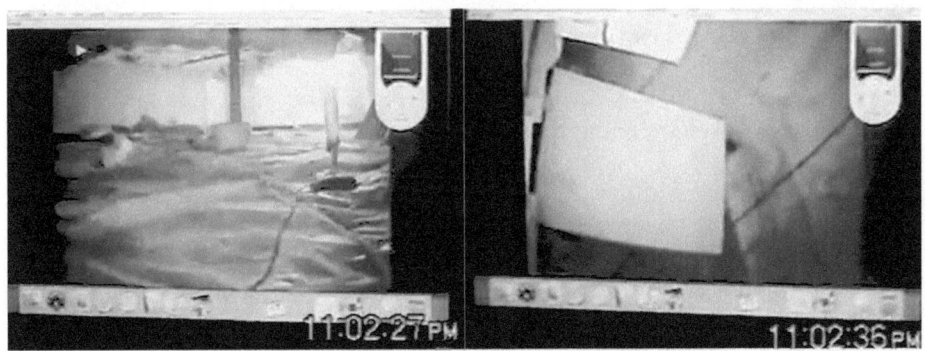

Figure 20.2. Screen captures from the crawlspace before (left) and after the camera tipped. Photos: RMP archives.

We rewound the footage to see just what had been happening while we were away and found that we'd recorded footage of the camera first jerking slightly as if bumped and then turning completely on its side. When we went down to check the camera itself, we indeed found the entire tripod had been knocked over.

Nothing on the video showed the culprit.

Several ideas came to mind. First of course was that a gust of wind might have knocked it over, but we quickly dismissed that idea. Not only was the camera in a crawlspace with the door closed, but it was a relatively low-wind day and the footage doesn't reveal any dust or other particulate matter (which was plentiful in the crawlspace) blowing at the kind of speeds that would be necessary to overturn the camera. The tripod we were using wasn't a cheap discount model but one intended for use in field photography by professionals, so it had some weight to it. Further, because the ceiling of the crawlspace was so low, the tripod was set with the legs spread wide. This significantly lowered the center of gravity, meaning it required a substantial force to knock it over. It wasn't going to fall just from a slight breeze or floor vibration.

Maybe a critter knocked it over? We thought of that as well. The family dog seemed like an easy scapegoat, but he was with us and the family in the back yard, so we knew it couldn't have been him. No other pets lived in the house. The possibility of another critter can't be entirely dismissed. Perhaps a rat or racoon or something managed to get into the crawlspace and knocked the camera over. While we can't rule that out completely, we don't think it's a great explanation. For one thing, we never actually found such a critter (and we certainly did look for one). The video doesn't show any other movement in the crawlspace, the majority of which was in the camera's view before it got knocked over, and we didn't find any place down there where vermin might be hiding. Nor did subsequent footage captured during the remainder of our five-day investigation capture any sight of any kind of critter in the crawlspace. So if it was an animal, it must have been one that found its way into and out of the crawlspace just within that short window of time and didn't come back for the next several days. That doesn't seem likely.

Another idea that came to mind was the cable leading from the camera back to our monitoring station. It was a long cable to be able to reach so far, and we thought maybe someone had accidentally tripped over it and the tugging caused the camera to fall. This explanation, too, we quickly ruled out. After checking that the cord was still where we'd left it tucked behind a large trunk and then jammed in the crawlspace door—it was—we then gave it a few firm tugs from outside and found it was secure enough we couldn't get the camera to fall over by this method.

We never figured out why the camera fell. After doing our due diligence, we set it back up and continued recording for the duration of our investigation. From that point on, it stayed in place and recorded nothing of any interest.

Point of interest number two occurred on another evening. When Oswald and Pauline had Quentin downstairs for a while, one of our team members took the opportunity to spend some quiet time in the child's bedroom to see if anything might happen. At some point, he fell asleep in the dark room, but the entire scene was captured on our infrared video cameras.

For a long time, nothing happens in the video, but there is one moment that captured something we can't quite explain. A sort of thump or tapping sound can be heard on the camera. This awakened our sleeping team member and though the sound's point of origin isn't clear on the monophonic recording, our team member said it sounded liked it came from the direction of the crib. While he's in the process of waking up, the video shows a shadow moving rapidly across the open doorway behind him. He didn't notice this part in person. Shortly thereafter, other people re-entered the room, the lights came back on, and nothing else unusual happened.

What can we say about this moving shadow? First, we checked to see who else might have been moving around that might have wandered past the door. The shadow did appear to be roughly human-sized, and though it seemed to move faster than a person merely walking down the hall, the fact that the hallway behind the door was in deep darkness means a person running past might have looked a lot like that shadow did. But no one else was moving around at the time. The shadow was too large to have belonged to the family dog, who was also not present at the time anyway. Our team member wasn't moving, so nothing he did could have cast a moving shadow. Unfortunately because the room was dark, the video footage is of insufficient quality to learn much more than that. It remains one of those intriguing little puzzles we're not quite prepared (yet) to call genuine evidence of the paranormal, but for which we also can't come up with a conclusive explanation.

Five long nights later, we brought our investigation to a close. We were happy to have been able to spend some extra time in the house because we did witness those couple of strange phenomena, and we certainly wanted to help the family figure out what was actually happening. Nothing else of note happened while we were present, though, so we eventually had to pack up our things.

Wrapping up, we had a long conversation with the family. We shared what we had experienced and what we had done to try to explain things but at the end of the day we had to leave them with "we don't know" as the only honest answer. Though we had to go home for now, we also said we'd be happy to conduct follow-up investigations if anything new happened.

Then we had the more difficult conversation, touching on the psychological issues implicated in the case. Emphasizing that we were speaking without judgement, we raised the questions of the family's fighting and the untidy state of the house and suggested they should seek counseling and make a concerted effort to start getting things back in order.

Our argument to them was the same as the argument we'll make here: we don't know whether the house is haunted or not. And we don't know whether uncleanliness necessarily has anything to do with it. But we've found that often simply putting your life back in order is sufficient to make the paranormal phenomena stop.

Clean your house and the demons will go away—that's a phrase we've come to quite appreciate because it seems to apply whether one believes in the supernatural or not. If you do believe the house is haunted or demonically infested, there's plenty of lore in the supernatural literature to suggest that our own behaviors can attract or repel certain kinds of entities. In that analysis, the filthy house could be attracting some kind of negative entity. If you clean your house, the demon will go away. Conversely, if you don't believe in the supernatural, there's plenty of support in the psychological literature to suggest that the conditions under which you live can affect not only your well-being but also your perceptions. The alleged paranormal entities could be psychological manifestations of your own troubled mind. And if you clean your house, those demons tend to go away, too. Either way, the better you can do to improve your life, your family's relationships, and your living conditions, the less likely you will be plagued by these kinds of terrifying happenings.

Clients don't always take that kind of advice well, but we were pleased in this case that they understood our argument and agreed that they'd try to put things back in order. First, they'd work on cleaning up the house and then they'd seek some kind of counseling to mend their fraught relationship.

Two weeks later, we contacted Oswald and Pauline to follow up and see how they were doing. They said things had gotten much better, both in terms of diminishing paranormal phenomena and in terms of improving conditions in their lives. We congratulated them on their progress and insisted they should never hesitate to contact us if they ran into any more trouble. They never did.

21

The Light Ghost

Wrapping up our collection of private cases for this volume is this doozy of a case that brought us in to an active investigation after a former team botched their data collection, offended the clients, lost some of their equipment, and abandoned the entire thing without another word. One or two strange events did occur during our own investigation but the case stands even more as a case study of how the other team did everything wrong than a truly frightening ghost story in and of itself. If you're any kind of private investigation professional—paranormal or otherwise—you might want to pour yourself an adult beverage before you read this one.

Figure 21.1. Artist's rendition of the Light Ghost. Image: Spraycasso.

As we draw this section of private cases to a close, we thought it would be wise to end with one that most dramatically emphasizes some of the themes we've been discussing throughout this section. Because we weren't the first team on scene, we'll be able to spend some time contrasting our methods with another group which we won't name, but we will mention that it was the same group we encountered in one of the prior chapters, meaning we've spent a substantial amount of time cleaning up their messes. Believe it or not, though, these people are far from the worst we've encountered, but that's a story for another day and another volume.

This story takes place in a two story single-family home with three bedrooms, three bathrooms, a basement bathroom, an attached garage, forced air heating and an active sump pump in the basement. After we were brought in for our investigation, we checked the home's property records and determined that it was built in the late 1970s (we know the exact date but are keeping that private to further protect anonymity) and had been owned by about half a dozen residents over the years (again we know the exact number but omit it in the interest of discretion) including our clients. We didn't find any reports of major crimes or tragedies taking place at the property, but that's only as conclusive as our search into property records allowed.

Occupants of the house were a married couple who we'll call Rosie and Stephen along with Rosie's brother Tony, who lived in the basement.[1] They also kept two dogs as pets. When investigative activities began, first with another team and then a little bit later with Rocky Mountain Paranormal, they'd lived in the house less than a year.

Most of the information we have to report about this case's background came to us out of order and in snippets. For ease of comprehension, we're going to describe the background first, then discuss how we became involved and what we were able to discover.

Rosie was not only the individual who contacted the paranormal teams but also seemed to be the one most affected by the alleged paranormal phenomena. She said the family began experiencing a variety of paranormal activities the entire time they lived in the house. Rosie said she also had a long history of paranormal experiences in her prior homes. Something seemed different about this house, though, and the activity eventually reached the level at which she and her family felt the need to reach out for help.

---

1   An important note regarding how we've chosen to fictionalize these names seems to be in order. At one point in the investigation, because of comments made by the prior investigative team, the clients' cultural or ethnic background becomes relevant to this case. As part of our effort to avoid publishing identifying information, we have semi-randomly selected relatively common (at least in the United States) names from a list of baby names following the same selection strategy employed in other chapters. The clients' real names might have at least hinted at their cultural background and, though we will discuss that issue when it comes up, we're making every effort to do so in a way that does not provide any information that could lead to a reader's ability to identify the people involved.

When Stephen was away on business, Rosie said she heard the sound of footsteps throughout the entire home. On multiple occasions, all of the home's residents have reported hearing footsteps and disembodied voices on the main floor and in the basement. This even includes the two dogs who were often seen barking at "nothing" and biting at empty air.

The toilet in the master bedroom flushed repeatedly and refilled for an uncharacteristically long time before stopping without apparent reason. Unlike in our previous encounter with a Toilet Ghost (see Volume 1, Chapter 23), this family claimed they had a plumber examine the toilet before they were willing to consider it a paranormal claim. They said the plumber couldn't find anything wrong or unusual about it. That's an important detail because our first piece of advice to anyone with this sort of claim is to check for natural explanations before jumping to the ghostly conclusion. Perhaps a ghost could flush a toilet, but it's just as likely a leaky flapper valve causing it to flush at random intervals. Because a plumber checked it out, though we're not yet prepared to say conclusively that something supernatural was afoot, we do give the story a bit more credence.

More troubling, and in line with several of our other stories in this volume, the family reported that something about the house seemed to be damaging their relationships. Stephen and Rosie in particular said that in addition to being husband and wife they were always the best of friends. But since they moved into the house, their relationship turned to frustration and fighting. Things got so bad that Stephen would leave the house for brief periods. Instead of being a respite for both of them, though, Rosie said the paranormal activity in the house would worsen during Stephen's absences. We always note that it's impossible to determine the causation of these kinds of things. Could relationship troubles make people more prone to paranormal experience (either for psychological reasons or at least hypothetically by attracting a negative spirit)? Sure. Those kinds of reactions are well-documented. On the other hand, could moving into a place plagued by paranormal phenomena (whether or not the phenomena have natural explanations) be the kind of stressor that could strain a relationship? Absolutely. Such cases are also well-documented in both the paranormal and psychological literature.

It wasn't long before the family decided it was time to reach out for professional help. As mentioned, we weren't the first team they contacted. They sent an email to a local ghost hunting group who took a month to respond to the initial contact. When they wrote back they said they were still trying to figure out which of their members should be responsible for responding to incoming requests for help and so this family's email fell through the cracks for a while. Good job it wasn't a real emergency situation! Nevertheless, arrangements were made for this team to conduct an on-site investigation.

Often, we only learn about the results of other teams' investigations by what the clients are able to remember and report to us. In this case, though we did receive reports of that kind, we also came into possession of more direct records of the team's activities so we're able to much more accurately describe how they operated. We're going to go through some of their practices and add our own commentary along the way because it seems like a good opportunity to highlight what they did right and what we think they could have done a lot better.

Their initial interview was conducted at the same time as their investigation. That's reasonable enough. We personally prefer, when possible, to interview people in advance, and often at a neutral location, but there's nothing inherently wrong with conducting the initial "interrogation" on site. They then told the residents they should stay in the living room while the investigators did their work.

Such a strategy is both good and bad. On the positive side, it's a very good idea to not have everyone just running about the house and potentially contaminating data. Confining people to a single room means their location is known so there's never any question of who might have caused any anomaly that's later recorded. The disadvantage of this approach is that paranormal phenomena do not occur in a vacuum. They happen in real time and space while people are going about their lives. It could be that some of the family's activities might be relevant. When we investigate, we like to mix things up a bit. There are times when we want everything completely quiet so we can record without disturbance. But the rest of the time we tell the clients to do their best to go about their normal daily business as if we weren't even present. We can then adjust our protocol in real time to adapt to anything that may or may not happen.

With the family in the living room, the investigators started with an EVP session in which they walked throughout the house with audio recorders asking questions of the spirits in the hopes of recording a reply. After a little more than ten minutes of this more active EVP session, they decided to go "remote" and conduct a more passive EVP session in which they left the recorder running but vacated the house. That portion of the investigation lasted about half an hour and was recorded on two camcorders and one digital voice recorder. If you think we have some thoughts about EVP sessions, you're absolutely right. But we're going to save those for a little later in our commentary, after the case got fully handed over to us.

The team concluded their investigation and said they'd be in touch with the family after they'd had a chance to review their data. For us, this was entirely premature. It's possible to imagine a paranormal investigation lasting just a couple of hours. Some of ours, in fact, have been pretty short. But it seems to us like there was no good reason to leave so early. We're looking for unpredictable anomalies in this field, so we have to be dedicated enough to stick it out for as long as possible in the hopes that something interesting will happen or that we'll have enough opportunity to get to the bottom of the mystery. We don't want to criticize too harshly, though, particularly if we interpret this as only a preliminary investigation. If they planned to come back to follow up, the brevity of their activities might make some sense. Trouble is, they never came back.

Soon after the team left, the family discovered they'd left behind the digital voice recorder they'd been using for their EVP session. It was sitting on the fireplace mantle. They reached out to the investigators both to make arrangements to return the recorder and to follow up regarding whatever the team might have discovered. Recall that these are people who are frightened or curious enough about their own home that they'd invited ghost hunters in, so it's easy to see why they might be a bit anxious to hear back. The investigators told them they'd be back out to collect the recorder in about three weeks.

Weeks passed. A lot more than three of them. Still no contact from the ghost hunters. The family wrote to them again and were told someone would "eventually" be out to collect the recorder.

More weeks passed. Eventually the family wrote them again to see if there were yet any results from the investigation. The team told them there was no evidence, that they'd "felt" nothing during the investigation, and therefore the home was "not haunted." That was actually surprising to us because this was the same team that had been willing to consider some really flimsy evidence as proof positive of a haunting in past investigations. But even here, they weren't getting it right, albeit in the opposite direction.

Evaluation of evidence—in the paranormal and in general—is basically a balancing act. Statisticians describe two fundamental kinds of errors one can make. Type I errors, or false positives, are technically defined as rejection of the null hypothesis when it's actually true. To oversimplify in the paranormal context, that's the same as saying there's a ghost when there really isn't. Type II errors, or false negatives, by contrast, are failures to reject the null hypothesis even though it isn't true. In the paranormal context, that's like saying a place isn't haunted when it really is. Some statisticians even talk about Type III (or even Type IV) errors, which basically involve coming to the correct conclusion but for the wrong reasons. Though these are not widely accepted designations in the statistical testing literature, we think they can actually be useful in a field like the paranormal, because it's entirely possible this other team made a Type III error. If we assume the house was not really haunted (we're not saying that's the case, but assume for the sake of argument), their conclusion may have been correct; however, they were entirely unjustified in jumping to that conclusion on the basis of the evidence available to them.

Understandably, the family involved didn't particularly like that conclusion. It wasn't so much that they disagreed with the ultimate conclusion—though they certainly did think their house was haunted—as that they weren't satisfied by the response. In a follow-up letter they said they'd listened to the tape recorder and heard some sounds they couldn't identify and wanted to discuss the matter with the team to learn more about it. The only response they received was that the tape recorder didn't belong to them and they shouldn't listen to it.

Getting annoyed now, they also added that on this recording they'd heard members of the ghost hunting team make derogatory comments about the family's home decorations and ethnicity. They were annoyed by the unprofessionalism, offended by the comments they'd heard on the tape, and ready to talk to someone else.

They reached out to The Atlantic Paranormal Society (TAPS) of *Ghost Hunters* TV fame to see what could be done. Rosie's letter to TAPS described their experience with the prior team and reiterated the intensifying nature of their paranormal claims (as we've already described above). We'd worked with TAPS before (see Volume 1, Chapter 12), so because they weren't local and unavailable to help out directly, they wrote back to refer the case to us: "I need to see if Bryan from Rocky Mountain Paranormal could check your house out. He is in your area and is a great friend who also runs a top-notch group. I am cc'ing him on this email. As for the recorder the last group left. If they don't contact you back for it after the situation that happened, then I guess you could either give it to Bryan from Rocky Mountain Paranormal or keep it. I am sorry about the last group you had conducted and in no way endorse their behavior."

Thus did Rocky Mountain Paranormal get involved in a case that had already become so much messier than it really should have. We contacted Rosie by telephone for some initial discussions and eventually scheduled an investigation of our own. During the course of this investigation, the clients gave us the digital recorder the other group had left behind. Though we didn't actually get a chance to listen to the tape until after our own on-site investigation, we're going to discuss what it contained now just to close the book on what the other team accomplished before we move on to our own investigation.

The recording was taken on a digital audio recorder left in the home near the fireplace mantle. Its length is just over an hour. An EVP session starts at the beginning of the audio and lasts for the first eleven minutes. The "remote" session begins at that time and ends at just after thirty-five minutes, at which point the team reenters the home. Some unintelligible voices follow. The ghost hunting team appears to depart at forty-eight minutes, after which the clients seem to let the dogs back inside and turn on the television. These nondescript sounds continue until the end of the recording at exactly one hour, one minute, and twenty-one seconds.

First, let's tackle the question of derogatory comments made about the family's ethnicity. We did indeed hear the remarks in question, and we want to tread carefully about making any conclusions as to whether or not the family were right to be offended. Everyone has an opinion on these kinds of sensitive issues and it's not our place as paranormal investigators to let our own opinions affect our judgments, so we're not going to say what is or isn't—or what should or shouldn't be—offensive. Instead, we'll simply describe what we heard.

The comments in question were in regard to the home's décor. One member of the team noticed a large number of certain types of ethnic décor (they were specific but we're keeping things vague both to avoid giving offense to anyone and to help preserve the clients' anonymity) and asked another team member about it. The second team member responded first that it was just "feng shui."[2] As the conversation continued, this team member ascribed aspects of the décor to European and Eastern Indian influence. None of those were correct.

We don't think the comments were hatefully intended. Nor do we think the reason Rosie and her family found them offensive was because the team member incorrectly identified the ethnic influence of the home décor (though we can say their guesses were *way* off). Rather, we think what Rosie and family found offensive was that the décor came up at all in the context of a paranormal investigation. They seemed to be suggesting, though no one explicitly said so, that some sort of religious practice might have some bearing on the paranormal phenomena. This is a delicate subject because there absolutely are times in which the religious beliefs of a client become relevant to a paranormal investigation. In our own work, it's one of the standard questions we ask on initial interviews, because we want to make sure both that we understand the religious perspective people might have about the paranormal (for example, some religions are more

---

2   We saw the décor for ourselves. It had nothing to do with feng shui.

likely to ascribe anomalous activity to a demon and others to an ancestral spirit) and so we don't accidentally do something to violate someone's religious practices when we're guests in their own home. That's useful information to have for a variety of reasons.

But on the other hand, it's not appropriate to pass judgment on the religious beliefs of others unless there's very good reason to do so. We've heard of plenty of cases in which police or first responders (not necessarily paranormal investigators) have misunderstood religious altars in people's houses leading to unwarranted questions about devil-worship or membership in cults. As a general rule, we figure it's best to either ask up front if the question seems relevant or to simply ignore it.

Moving on to the paranormal investigative side of things, this recording does appear to contain some unusual or anomalous audio. Unfortunately, however, it also represents a case study in the incorrect acquisition and processing of data and therefore has little to no evidentiary value.

During the initial EVP session, there was insufficient control to obtain usable data. Rosie might have been offended by the ethnic commentary on the tape, but what struck us from a scientific perspective was *that* there was an off-topic conversation occurring during the EVP session. And it wasn't just limited to this one question about the home décor. At one point we even heard the investigators discussing computer software they'd recently purchased. The idea with an EVP session is to attempt to record ghost voices that might be detectable in recorded audio even if they can't be heard by the unaided human ear. There's no real science to suggest that should work, but it's common enough in the paranormal lore that we're not entirely uncomfortable with the practice…but only if it is conducted with scientific controls.

First, you need to give the supposed ghosts the opportunity to actually respond. You need clean audio. In this case, a ghost could have wandered directly up to the microphone and started reciting Shakespearean sonnets and we'd never know about it because the investigators themselves never shut up long enough for the spirits to get a word in!

Second, it matters how you evaluate EVP evidence (not that this team ever got to that stage because they left their recording behind). If you go into a place and start asking questions of the spirits, you're psychologically priming yourself to think that there is a spirit present. And the nature of your questions will set you up to expect certain responses. If you record some ambiguous audio, it can be almost impossible to avoid hearing what you expected to hear, rather than what you actually recorded, under those kinds of conditions. Therefore, we insist that EVP sessions should always ask their questions hypothetically. A good additional practice would be to ask questions with very specific—and known—answers, but to have to the audio analyzed and transcribed by an independent individual who doesn't know the question or answer. In that kind of blinded experimental design, you minimize the opportunity for psychological biases to affect your results. The questions asked by this particular team were fairly standard "EVP questions." They asked things like "when did you die" and tried to convince the spirit or spirits to make their presence known by making sounds or moving objects.

One thing they did absolutely correctly, though, was to audibly call out whenever they discharged a flash on one of their cameras. That way, had they ever listened back to the recording, they'd know that those sounds were caused by their own people and not by a ghost. Though we'd prefer if they'd done the different components of their investigation at different times rather than all at once, their effort to keep track of who was making noise should be commended.

The "remote" session is a good idea to obtain "cleaner" audio without the potential for these human contaminants, and much closer to our own investigative protocols, though we use different terminology and usually don't actually leave the location. Instead, we usually just hunker down at our base of operations and shut up while we watch and listen to all of our monitors. But that's just personal preference. The idea of vacating the house with the recorders left on is a good way to remove noise from the audio and just see what happens.

Indeed, the only audio of potentially paranormal interest on this tape occurs during this portion of the recording. Throughout this segment, it is possible to hear a variety of sounds that are difficult to explain. Knocking and shuffling sounds can be heard, as if someone were moving objects or furniture within the house. Some of the sounds are easy enough to identify. Windchime sounds were just the windchimes on the porch. A "warbling" kind of sound puzzled us for a minute but we later determined it was just the house's climate control system activating. But those sounds of knocking and furniture scraping along the floor aren't so easy to explain away.

Because dogs' barking can be heard on the tape, one possibility is that, despite the team's claim to have vacated the premises of all people and animals during the "remote" session, one or more dogs remained in the house and were responsible for the sounds in question. Maybe a dog pressed against a chair and slid it along the floor. However, comparing the sound of the dog barks during the remote session to those heard later in the recording, after people have returned to the house and let the dogs back inside suggests this is not the best explanation. The dog barks and footsteps are substantially louder and less muffled after the remote session than during.

Another possibility we considered was that of a deliberate hoax. Some of the sounds certainly at least raise this as a possibility, as some of the sounds seem suspiciously like footsteps, moving furniture, or doors opening or closing. While this remains an open possibility to account for the sounds on the audiotape, it is our opinion that it's not the most likely. Given that the audio recorder was left behind by the ghost hunting team, and given our interactions with this team in the past, it seems more likely that if they were to attempt to perpetuate a hoax, they would have recovered the audio and publicized it themselves, rather than leaving it for another team to discover later.[3]

Maybe a window was left open and wind blowing through the house could explain some of the sounds. That windchimes (located on the porch) can be clearly heard in the audio does lend some support to the idea that a window

---

3   A *smart* hoaxer might indeed leave it for someone else to make the discovery to dispel suspicion. But we also don't think the team in question are the sort of people to intentionally perpetuate a hoax, nor do we think they likely could have pulled it off without the family's knowledge.

might have been left open. And wind could certainly explain some of the sounds. Knocking, for instance, might just be an object blowing in the breeze. But we're skeptical that the wind could explain the sound of furniture scraping along the floor because if a wind were strong enough to move tables or chairs, the sound of the wind itself ought to also be clearly recorded, and it wasn't.

Ultimately, because Rocky Mountain Paranormal personnel were not present during the setup of this so-called remote session, we are unable to verify the conditions under which the audio was recorded, and the sounds remain unexplained—and likely unexplainable. The thought occurred to us of trying to recreate the experiment under stricter scientific controls but we were ultimately unable to do so. As such, this audio joins a growing collection of mysterious recordings we can't quite explain.

But of course we didn't just get the audio recorder and pass judgment on the other team (as fun as passing judgment on people can be from time to time). We had to conduct our own investigation into the property and their various paranormal claims. We arrived for a one-evening investigation at around 7:00 p.m. on the day we'd arranged with the family and immediately began setting up our equipment and documenting the condition of the property.

Setup for this investigation was pretty straightforward and typical. We placed video cameras at the bottom of the main staircase looking up to the second floor, in the kitchen looking toward the living room (though at 1:40 a.m., we moved this one to the top of the stairs), in the living room looking toward the kitchen, in the master bedroom facing the hallway door, in the basement living room, and in the main floor living room. Microphones went in the main floor living room, at the top of the main stairs, at the bottom of the basement stairs, and on the kitchen countertop. Finally, remote thermometers were placed at the top of the main stairs, at the bottom of the basement stairs, and in our "base camp" monitoring location, where we spent the majority of the evening.

Also as typical, we took hourly EMF readings from thirteen locations throughout all three floors of the house. The only EMF anomaly we detected was an elevated reading at one location in the stairway to the second floor. Throughout the evening the EMF level hadn't deviated from the baseline of one milligauss, but at about 10:45 p.m., it jumped to two milligauss. That's still not a terribly high reading, and we were able to determine the elevation was caused by the heater turning on.

Temperatures throughout the evening did not significantly fluctuate. We didn't find any spikes or major drops in temperature and because the house was climate controlled by a heater and air conditioning unit, interior temperatures never fluctuated by more than a couple degrees throughout the evening.

For their part, we told Rosie, Stephen, and Tony to just go about their usual nightly routine and pretend we weren't even there (though of course we told them exactly where all our cameras were both so they wouldn't accidentally disturb one and so they didn't need to worry about us accidentally recording anything embarrassing). They did so, and a few hours into the investigation went to bed. For our own part, except for our hourly stretch and EMF recording, we stayed put at our monitoring location.

The only moment of any particular excitement was at 12:17 a.m. when the bathroom light at the top of the second floor stairs suddenly turned off. We quickly confirmed that no one was up and about. Then at 12:35 a.m., the same light turned itself back on just as suddenly as it had turned of. Still, no one was in the room or anywhere near it at the time. We stopped our usual monitoring protocol, checked that the room wasn't occupied, and fiddled with the light switch for a while to see if we could figure anything out. We were unable to find any apparent cause, and at 1:40 a.m., we moved another camera to this location to monitor it more closely but no other anomalies occurred.

Figure 21.2. The light that turned itself off (left) and on. Photos: RMP archives.

Other than the light turning on and off, we didn't detect anything strange on video. Audio turned out to be all but useless in this case because the family's dogs were in the house all night and, probably a bit confused by all the extra people, were a bit on the noisy side.

We left the following morning without having come to any significant conclusions. We were curious about the light's unusual behavior, of course. One possible explanation we thought of was potential faulty wiring but we didn't have the equipment with us at the time to test for that. And as much as we wanted to go back to follow up, we were never able to do so because the family had decided to move due to a combination of their haunting experiences and a new job opportunity.

Though our own investigation didn't yield much fruit, we're still pretty happy with the outcome of this case. As much as we wanted to get back in and figure out what happened with the light and to see if we could document any more of the family's claimed phenomena, we're certainly happy that at least in this case the family's decision to flee the house wasn't a financially-devastating one. The case also gave us the rare opportunity to directly examine another group's data and to contrast their methods with our own. But at the end of the day, it's still a mystery and the only reason we consider this case closed is because we no longer have access to the location.

PART THREE:
Media Analyses and Other Activities

Activities of the Rocky Mountain Paranormal Research Society are not and have never been limited to investigations alone. While those do take the bulk of our time and represent the flagship of our work, we have our hands in a lot of other things as well. We now turn our attention to those cases that are somehow different from the standard investigations (whether public or private) we've been discussing so far.

In many ways, this third part of the book (and the corresponding parts in each volume of the series) represents something of a grab bag of different and often unrelated activities, but most (not all) of them fall into a couple of categories.

First are the media analyses, and those can be subdivided into our analyses of media that has been published and received enough attention to merit our commentary and our analyses of photographs or videos sent to us along with an invitation to perform an investigation but which never develop into a full-fledged investigative activity (typically because our analysis of the photo or video is sufficient to explain the phenomenon in question).

Then there are the experiments or other forms of academic research. From time to time, independent of any specific investigation, we perform a variety of experiments whose results may illuminate some aspect of paranormal lore or phenomena. We've done several such experiments in the past, one of which was included in Volume 1, though our experimental "department," so to speak, is beginning to pick up more steam in recent years, so later volumes will likely contain more of those cases.

Sometimes we also get involved in cases that aren't so much paranormal investigations as they are public services or civil or political actions. Clarity is in order here. Politically, Rocky Mountain Paranormal is committed to neutrality. We're an investigative and educational organization. Members through the years have been Democrats, Republicans, and other. We don't espouse any particular political or economic philosophy, we require no political test for membership, and we're committed to remaining non-partisan in our work. However, every once in a while, a political issue emerges that is specifically related to the paranormal. When that happens, we feel compelled to take action to ensure that everyone involved has the highest quality information available and behaves ethically. One such case will be included in this volume.

Finally, sometimes there's a case that just doesn't seem to fit into one of the other standard classification of investigations at either public venues or private residences. Even if we stretch the meaning of those phrases a bit, some of the things in which we involve ourselves just seem to belong more properly in this kind of a grab bag part of the case files books.

Though this part is relegated to the back of the book, don't be fooled into thinking it's less important than the others. Every case in which we participate offers something interesting and educational, and we hope you enjoy reading about some of our less typical activities here.

ESSAY
The Paranormal and the Press

We have something of a love/hate relationship with the press at times. Some of our members at various points in their lives have seriously considered a career in journalism. Even though we didn't ultimately go down that road in the traditional sense, we consider some of our work to be at least journalism-adjacent. We may not work for a newspaper or (thank God) a cable news network, but we are still conducting investigative activities which we report to the interested public both in these books and in a variety of other media.

Paranormal aside, journalism (and the press generally) are fundamental to everything we do. Without them, none of us would be informed of current events, none of us could learn the important historic or scientific facts that keep our little gray cells excited. Indeed, civilization itself depends on this exchange of ideas and information. At their best, journalists are the fearless defenders of truth in a world filled with lies. And even when we're not talking about news media or investigative journalism, the people working in the media at large are those who keep us both informed and entertained.

But that's at their best, and they're not always at their best. Obviously this isn't the place to make any political arguments about the state of journalism. We're pretty sure we don't even agree amongst ourselves on some of those points of contention. Politics aside, though, we don't think anyone can seriously disagree that not everything that gets published in the newspapers or covered by the talking heads on the morning news program is accurate.

One of our members recalls a time when an acquaintance of his got written up in the newspaper after being victim of a crime. The paper managed not only to get his name wrong but to turn him from a forty-something male into a twenty-something female of an entirely different ethnicity. Surely that kind of error is not representative of most of the press, but it does raise an important question: if they get the facts wrong about a matter you know something about, why do you trust their information when it comes to a matter you don't know anything about?

Our purpose here isn't to criticize the media per se, but to argue for improved media literacy (whatever that means, and people will disagree about how to go about that as well).

Returning to the paranormal, we've noticed there is often an ebb and flow to media coverage. There's always some baseline level of interest in these matters reflected in programming specifically dedicated to the topic. Ghost hunting shows may not be quite as popular now as they were at what seemed to be their peak a few years ago, but they still reach a wide enough audience to keep getting made. Entire magazines containing nothing but paranormal stories and related articles are still published regularly. And the more mainstream news does still take an interest from time to time, either when something big happens in the paranormal world, when they have a particularly slow news day and need some filler, or reliably every October in preparation for Halloween.

Because we do what we do, we've not only paid close attention to how the media treat paranormal topics, but we've had a front-row (and often behind the scenes) look at how such programming gets made, so we wanted to take this opportunity to simply offer a few general observations that hopefully will inform your paranormal media consumption in the future.

Observation number one: almost everyone has a bias when it comes to this subject. Paranormal publications tend to fall pretty firmly into factions of believers and disbelievers. Of groups and publications specifically dedicated to the paranormal, we can't really think of any other than ourselves who maintain our strict neutrality. We're sure there must be some out there, and even those that don't maintain neutrality often at least pay lip service to our kind of philosophy.[1] This isn't to criticize those publications. We're devoted readers of plenty of them on both sides of the issue. But you just need to know that if you pick up a copy of *The Fortean Times*, you're getting a more sympathetic view of the paranormal and if you pick up *The Skeptical Inquirer* you're likely to find more debunking.

Even in mainstream press without a specific institutional bias for or against the paranormal, the articles tend to come from one perspective or the other. A local newspaper or television "puff piece" about a local supposed haunt is unlikely to take the time to bring in a forensic investigator to seriously question the lore. And on the other hand, a major newspaper's science page isn't likely to spend a lot of time studying the paranormal lore before trying to find a natural explanation.

Even if they keep their conclusions neutral, they're reporting from a particular perspective. Knowing that perspective will help the reader or viewer detect any mistakes of fact they might have made. More importantly, in the absence of errors of fact, it will also help the reader detect any potential omissions from the story. After all, it's possible to tell nothing but the truth without telling the whole truth.

Observation number two: reality television usually isn't. We're not here to pee on anybody else's parade. We love the paranormal reality shows, too. Some-

---

1   That may be unfair. Some of them go beyond mere lip service and actually do a pretty good job of treating any given case with as much methodological neutrality as they can muster. But they're still generally approaching the case with a particular bias in mind.

times at the end of a long day we're more than happy to watch some people go into a haunted place and scare the bejeezus out of themselves. But while they put on a good show of doing the kinds of things we do, that's not what's really happening. We've been behind the scenes on enough of those shows—and studied a bunch more from a distance—and we can confidently say that those programs are more interested in making an entertaining TV show than in reporting what actually happened.

In one instance, we know of a show that started out as more of a documentary kind of program and ended up as more of what we now recognize as a stereotypical paranormal reality show. The transition occurred following a specific episode. Something strange happened. It wasn't faked, but it happened and was caught on camera. Ratings for that episode spiked, quite understandably. Equally understandably the producers told the show's hosts that more of that thing needed to start happening all of the time because it made for better television. Fair enough, in the proper context. Television and science don't have the same goals. But the end result on that show and on a lot of other ones is that if nothing exciting seems to be happening, the cast or crew will make something happen. We've heard of at least one show in which the onscreen talent is completely honest, but some members of the crew carefully arrange faked paranormal phenomena just to get the stars' (legitimate) reactions on screen. We're not sure if that's true or not, but we're more than willing to believe it could be.

This can also be done without any fakery. Assume for a moment that the investigators on screen are completely honest and that no one is setting up any kind of hoax behind the scenes. The investigators might spend at least several hours and often several days working at an investigation. With multiple camera angles and all the other recording media they might have, this could easily total many hundreds of hours of footage. Editors can take the opportunity to show only the one or two moments in which something exciting happened and leave the rest on the cutting room floor. From a filmmaking perspective that's not only acceptable but the right thing to do. But it can also lead to false impressions about how haunted a place really is.

So the important lesson to take away is that a show made for entertainment, even if everyone is honest about it, does not give you the same perception as a documentary or (better yet) a scientific or journalistic report.

Observation number three: fiction accounts for a lot. So far, we've been talking about paranormal non-fiction, whether legitimate or hoaxed. But paranormal fiction also influences the culture, the culture influences the media, and the media influences belief. There is a small but growing literature on this subject drawing on interdisciplinary work in psychology, sociology, communications, and media theory. It's important to understand that the relationship between belief and media is complicated and even when one finds a strong correlation (as between those who believe in the paranormal and those who watch paranormal reality television), it can be difficult to determine the direction of causality. Do people watch these programs because they believe in ghosts or do they believe in ghosts because they watch these programs?

Unfortunately a thorough review of the relevant literature, while fascinating, is well beyond the scope of this book (if you think our books are long now, imagine how massive they'd be if we really pulled out all the stops on the academic side of things). However, we would like to highlight one recent (2020) master's thesis that attempted to separate paranormal belief by category (e.g, angels, demons, ghosts, UFOs, etc.) and media consumption by type (animated fictional, live action fictional, and reality or documentary) to determine whether the degree of realism in media had any effect on the corresponding belief in the relevant paranormal entity and whether this differed by type of entity. The results were complicated (and themselves beyond the scope of this essay) but the gist was that the type of entity did matter (in fact, media consumption was negatively associated with belief in the reality of angels and demons across the board) and, more importantly, the level of reality mattered a lot, with more realistic programming being more associated with belief in each type of entity.[2]

Combined with our own experiences working with a variety of people ranging from paranormal investigators (from both the skeptical and believers' sides) to our clients to authors and filmmakers and curiosity seekers and more, results like these show that it's difficult to predict why people believe as they do, but also that the media they consume tends to have a noted effect.

Fiction also affects, independently of any effects if may have on the prevalence of belief, the details of paranormal claims. Skeptical researcher Benjamin Radford spent several years tracking reports of the chupacabra (a cryptid most commonly reported in Puerto Rico and Mexico said to vampirically consume the blood of goats and other livestock) and, though we don't fully endorse all of his conclusions, he did convincingly determine that at least some of the chupacabra sightings were influenced by the appearance of the creature Sil in the 1995 movie *Species*.[3] On the subject of aliens and UFOs, art historian John Moffitt has also argued—again fairly convincingly—that the changing appearance of reported aliens over the years closely reflects the changing appearance of aliens in mass media and popular culture.[4] We can't speak for the authors we've cited but we at least are not trying to argue these creatures—chupacabras, aliens, or any other kind of paranormal creature—don't exist. We're simply pointing out that what people believe about them is influenced not only by non-fiction media but by the pop culture memes that find their way into the public consciousness.

Observation number four: sensationalism sells. It's true of the media in general and it's certainly true when it comes to the paranormal. Whenever someone reports on a paranormal story, there's going to be some baseline of interest in the report and it's going to come from a mix of believers and skeptics who will

---

2   Halsey, M. (2020). "The influence of The Type of Supernatural and Paranormal Media on The Belief in The
Various Forms of Supernatural and Paranormal Phenomena." Northern Illinois University: Graduate Research Theses &
Dissertations. 7087. <https://huskiecommons.lib.niu.edu/allgraduate-thesesdissertations/7087> (accessed June 22, 2024).
3   Radford, B. (2011). *Tracking the Chupacabra: The Vampire Beast in Fact, Fiction, and Folklore*. Albuquerque, NM: University of New Mexico Press.
4   Moffitt, J. F. (2003). *Picturing Extraterrestrials: Alien Images in Modern Culture*. Amherst, NY: Prometheus Books.

each go off and either praise or condemn the article depending on how closely it aligns with their own ideas. But in the middle are all those people who don't really have a dog in that fight and whose interest in the paranormal fluctuates according to how interesting any given story might be. Therefore, there's a strong incentive toward sensationalizing these stories. And that doesn't necessarily mean biasing them in favor of the believers. Skeptics, too, have their own forms of sensational reporting, and sometimes a sensational element can be emphasized without taking any side on the paranormal claim.

Good examples can be found throughout the paranormal literature, but on that last point we'll call your attention to the "evil undertaker" of Denver's Cheesman Park (see Volume 1, Chapter 5 for the whole story). Though it turns out he was just an honest undertaker doing a dirty job nobody else wanted to do, he's been reported in paranormal and mainstream sources alike as a ghoulish character who mishandled bodies for profit. We've been working to restore his name for well over a decade at this point but we still see the "evil undertaker" story reported in newspaper articles in a variety of Denver-area publications, usually every Halloween season like clockwork. They can maintain their neutrality about the various ghost stories of Cheesman Park but still get their over the top sensational story in. The tragic thing there is that the true story is sensational enough without having to dishonor the memory of Undertaker McGovern.

Sensational reporting might also explain the reporting of the United States Congressional hearings on alleged UFOs in 2023. Our inner conspiracy theorists are asking what they were trying to distract the people from by trotting out an alien story, but even without going down that particular rabbit hole, it's easy to see a discrepancy between what the hearings actually revealed (some hoaxes, some genuinely unidentified aerial phenomena, and a lot of politics) and what the press reporting led a lot of people to think the hearings revealed (that the United States Congress admitted to having discovered aliens).

Observation number five: balanced reporting is difficult and not always desirable. We get it. It's hard to actually be neutral on a story. Balance between the believers and the skeptics is a difficult thing to achieve. Admirable, but difficult. Both in personal experience and in our conversations with others, we've seen plenty of times in which a news report is planning on taking a particular view of some event but brings on either a "token skeptic" or a "token believer" to give some false sense of balance by pretending to represent the other side. Often those people are either made fun of, given insufficient opportunity to speak, or quoted so far out of context as to falsely report their meaning.

More well-meaning producers might legitimately try to balance their program by having a legitimate debate. But unless they have hours to dedicate to the program, giving each side ample time to explain their often-complicated perspectives, they're probably not going to get to the heart of the matter. Equal time in a formalized debate is not the same thing as legitimately giving people equal opportunity to explain themselves.

We've come to the conclusion that sometimes balanced reporting isn't even necessary. If the case has an obviously correct solution, balanced reporting is tantamount to creating the false notion that there's a debate when there really isn't. Better would be *honest* reporting. And that's what we always try to do in our own work. We absolutely have a perspective on some of these issues, despite our overall commitment to neutrality on the "paranormal question" in general, but we're up front and honest about what that perspective is, and so people who listen to our lectures or read our books, we hope, are able to make whatever mental adjustments are necessary to account for whatever bias we might have.

Observation number six: most mainstream media don't care whether a paranormal claim is true or not. This is kind of an offshoot of our previous two observations. Unless you're reading a publication with a particular interest in the paranormal (whether positive or negative), it's probably a safe assumption that the article's author and editors and publishers don't really care whether it's true or not. They may care to the extent that they don't want to report anything false, they may care whether one side is more likely to sell newspapers than the other, but they probably don't really have a deep interest in the particular question at the heart of the story. When this occurs, you're not necessarily looking for a bias, but rather you're looking for missing information.

Someone who has dedicated his or her life to a particular question probably has a lot of experience and knowledge. Those people will know all the little facts that might affect a story, but they might also have a bias in terms of which facts they choose to present. On the other hand, someone whose interest in the case begins and ends with the authorship of a single article may not have a bias but might also not have the foggiest idea where to dig up those elusive key facts that might change the whole thing. So pay attention to which kind of article you're reading.

At the end of the day, we're not telling you to distrust the media. Rather, we're encouraging you to understand the media more completely so you're mentally prepared to recognize what kind of article you're reading and to adjust how much credence you give it according to a wide variety of factors. Because when it comes right down to it, all of us are influenced by the media we consume. We don't start as blank slates, but neither are we immune to the influence of the world around us, and especially in modern times, much of that world comes to us through the increasingly diverse forms of media we consume.

Tread carefully. Be skeptical of everyone and everything—including the other skeptics themselves—and do your best to seek out high quality alternate perspectives. When in doubt, talk to some experts (preferably some from multiple sides of a debate), read extensively to educate yourself, and only then try to form the most educated and informed opinion you can. And then treat even that as tentative and changeable. In the meantime, if you're looking for some media you can trust about the paranormal, our own stuff (including both our investigative reports as you've already read and our analyses of media from other sources, some of which you're about to read) is pretty good.[5]

5　We kid. Kind of.

22
The Photogenic Moth

Many of our cases involve simply analyzing media that's sent to us for review from people's own camcorders or home surveillance equipment. When we initially planned to write these books, we thought of including a chapter in each volume simply reporting an assortment of photo and video analyses that never progressed to an on-site investigation. We abandoned that idea because there were simply too many low-quality images to be able to publish, others couldn't be cropped to avoid revealing potentially identifying elements of the clients' homes, and for a variety of other reasons. But in this chapter we're going to present three cases that came to us within period of a few weeks from three entirely different sources, all of which ended up having very similar explanations.

Figure 22.1. The Photogenic Moth. Photo: from video provided by the client.

None of the three cases we're presenting in this chapter really deserve to be called full cases on their own. They're simple and quick media analyses. We've collected them into a single chapter both because they're thematically related and because it makes the most sense to group them into one unit than to fill our book with excessive numbers of tiny chapters. It is important for us to include this chapter, though, because these miniature cases, such as they are, are representative of a growing collection of literally hundreds (perhaps thousands by now; we're still working on cataloguing and counting all of them) of similar activities.

You'd be forgiven for thinking otherwise given how many of them we've done, but our actual on-site investigations represent a minority of our activities. Most of the time, we're sent emails with some combination of photos, videos, and stories and asked for our interpretation. When appropriate, we schedule to meet the people on site and go for the whole shebang. More commonly, we're able to review the photos or videos and supply an appropriate interpretation right from the comfort of our own homes.

The first of these cases has earned its place in our casebooks under the title "Cats Will Be Cats." We were contacted by an individual who wanted to send us some video and get our opinion about what kind of paranormal entity might be bothering his cats. The felines had been acting strangely, particularly at night. One Sunday evening one of the cats kicked over her scratching post "in absolute fear," according to our client, and then "something flew down at her."

We agreed to review the videos and they arrived in our inbox a day or so later along with a lengthy explanation going into detail about the client's history in the home, the cats' entire life stories[1], and descriptions of the conditions under which the attached recordings were taken. In brief, the client said one of the cats in particular always seemed to be afraid of something and to spend her time chasing imaginary entities. The client himself reported often having trouble sleeping and "sensing" something unseen in his home. A local psychic (as usual, we do know who but we're not saying) confirmed for him that there were "spirits watching over" him in and around his house. To investigate further, he installed surveillance cameras in his own home to see what the cats were up to when he wasn't around.

The remainder of the email contains a description of each of the videos he sent us. Still images captured from the videos wouldn't aid the reader in understanding, so we'll do our best simply to describe what each videos shows.

Video 1: A cat jumps from an upper-floor landing to a sofa on the first floor and then walks away.

Video 2: A cat walks slowly through a room. When the cat leaves the frame, a glowing light appears and moves in the cat's direction. The cat then runs back through the camera's view in the opposite direction.

Video 3: A cat kicks over her scratching post then runs out of the camera's view, seemingly startled by a floating "entity" that's briefly seen moving through the camera's field of view.

---

1   We never expected to add "feline biographer" to any of our resumes. This is a strange line of work we've chosen for ourselves.

Video 4: Two cats and one human female are in the camera's view. A similar glowing light is briefly seen passing through, but the cats seem not to take notice.

Video 5: A cat walks in circles around the scratching post, seemingly looking for something.

The client described the objects floating through the camera's frame on several of those videos as "glowing lights," "things," and looking like "shooting stars." According to his report, the house normally has a "peaceful vibe," so he wasn't exactly scared of these phenomena but just wanted some information about what was happening and whether or not it was something he should be worried about.

We reviewed the videos and didn't see anything that looked paranormal to us. The first thing we noticed was that because of the low-light setting, the camera used infrared to illuminate the scene. Cameras of this type can make non-luminescent objects look like they're glowing. And indeed, that's exactly what was happening here. The "things" that had been floating through the camera view on several occasions were nothing but insects flying through the house.

In the video in which the cat knocked over the scratching post, we determined the object—we're pretty sure it was an insect—wasn't within the cat's field of view at the time, so the cat's behavior doesn't appear to have been connected to the floating object. However, that does leave us with the question of why the cats seemed to be acting strangely. We came up with three explanations. First, cats are natural predators and if they caught sight of a moth or some other insect, it's quite possible they would have attempted to chase it. Second, healthy animals often exhibit a behavior known to veterinarians and animal trainers as the "zoomies" in which they simply run around for no other apparent reason than simply because it's fun or to burn off some excess energy. And finally... they're cats. They're known to be strange and alternately lazy and excited, aloof and affectionate, and so on.

We figured this was most likely case closed but suggested if he wanted to be more certain, he could set up another camera or two watching the scenes from different angles and that, assuming our theory was right, he'd catch a better view of the insects and what the cats were up to. He wrote back to say he hadn't thought it might be insects because he rarely finds them in his home but would follow our advice and check his video to see if anything else happened. He didn't write back for more assistance, so we assume our explanation ultimately carried the day.

The next case, which we lovingly call "The Photogenic Moth," came to our inbox shortly thereafter. The client asked what we would charge to review some bizarre footage and lend our opinion. Not a cent, we said, and he responded with eleven short videos from his surveillance cameras showing luminescent streaks fluttering past the camera's frame.

The videos were taken on several cameras from several locations throughout his house and garage. A representative sample, cropped to remove any potentially-identifying home décor, is presented in the figure at this chapter's beginning. As you can see, it seems to show some kind of luminescent object. When viewed in motion, it seems to elongate and stretch across the screen as it flitters by rather quickly.

It was pretty easy to solve. If you look closely at the photograph, you can clearly see wings and antennae. This was a moth. As in the previous case, the reflection from the infrared camera made it appear luminescent even though moths in real life (particularly the miller moths common to Colorado) are dull in color.

Other videos from this client showed similar things. Some of them contained "orbs," which as we discussed in our previous book (Volume 1, Chapter 24) are readily explainable as dust, moisture, pollen, or other particulate matter reflecting light back to the camera. A few others appeared to contain a spider and its web, again luminescent for the same reason and out of focus due to their proximity to the camera.

Our favorite part of this case was the very end of one of the videos, and the reason we refer to this moth as photogenic. In one of the clips, the moth flies directly into the camera, producing an audible "thunk" upon impact. Over the years, we've examined more paranormal videos than we know what to do with. But this was the only time a "ghost" actually thumped into the camera. It's delightful and we wish you could hear it. Ask us at one of our public lectures sometime if you ever run into us and we'll play it for you. This was indeed the most camera-ready moth we've ever seen.[2]

Analysis didn't take us too long in this case. Just freeze-framing some of the footage clearly revealed the moth's anatomical structures (which, in the client's defense, were impossible to see when playing the video at full speed). So we wrote back quickly with our explanation. He took it well, thanked us for the explanation, and had no trouble accepting what we'd said.

That was refreshing for us because we never quite know what kind of reaction we'll get from someone when we issue a report. Some of them are relieved that it's not actually something they need to worry about. Others were merely curious and thank us for the explanation, as happened here. Some people, though, get so invested in their paranormal interpretation of the story that they get angry when our explanation doesn't match their expectations. We always try to clarify we're not saying there's no such thing as…whatever entity they thought it was. We're not even saying their house isn't haunted. We're just saying this one particular video happens to be a moth or a spider or a piece of dust or whatever it may be. That message resonates sometimes and sometimes they just stop writing altogether, presumably still angry about our conclusions.

The final case to come in this little batch of three seems to have been one of this latter group of people and is one we've named "Definitely an Insect." And the title rather gives away the game. We were sent two photographs via email. The body of the email just asked for our opinion about the two photographs, and the subject line read "Possible, Definitely an Entity."

Both photographs show the interior of the subject's home (with enough personal belongings and décor that we're uncomfortable reproducing them in print because it would be easily identifiable to anyone familiar with the house). They're taken from a motion activated camera from approximately the same angle. One shows the room during the day and we couldn't see anything in that one

---

2   Miller moths have a life expectancy of just a couple weeks, so RIP little guy. Your legacy lives on.

meant to be paranormal at all, so we assume it was included just to show what the room looked like. The other was a night view with the same kind of luminescent "streak" across the image we've been discussing.

As with the previous case, the freeze frame image clearly shows the antennae, so this is obviously a moth. We wrote back with those conclusions and the client never responded to us. It could be that we simply answered his question and he went happily on his way, but we have a feeling he might have been frustrated that his image wasn't, in fact, something supernatural.

Understandable. We get it. Every time we get a new case, whether it's a full-fledged on-site investigation or just one of these photo or video analyses, we're genuinely hoping it'll be the one in which we can finally say "we've done it" and found genuine incontrovertible proof of something supernatural. But we just have to go where the evidence leads, and the majority of the photographic or videographic evidence we've seen (both around the media and in our own casework) simply don't provide that kind of proof. At least not so far. Keep sending them and we'll keep trying.

23
The Tea Ghost

This brief chapter deals with an issue we hinted at in Chapter 6—the experimental recreation of paranormal footage. In this case, video came to our attention showing a box of tea packets floating in mid air. Though we didn't investigate on-site, we set about to determine what it would take to produce equivalent footage of our own. We might have gone a bit overboard.

Figure 23.1. Our re-creation of the Tea Ghost video. Photo: RMP archives.

The case of the tea-stealing poltergeist came to our attention a little more than a decade before the publication of this book when it started receiving coverage on a variety of news outlets. Allegations of genuine paranormal activity caught on tape often attract media attention, particularly if it's otherwise a slow news day (but even if not a slow news day, it often brings a much-desired bit of variety to a news program). Unlike so many others that involve such low quality footage that it's almost impossible to tell what's really going on, this one, while certainly not filmed in cinematic high definition, provided a clear video and an easy to understand phenomenon, so it quickly captured the public attention.

Both for copyright reasons and because it's impossible to recreate film in print[1] we can't show the original video, but we can describe it in detail and direct you to where you can find it (at least as of this writing) archived on the Internet.[2] The video appears to be a supermarket surveillance video. A man is standing in an aisle facing shelves on the left of the frame. From shelves behind him and to his right a red or orange box slowly removes itself from the shelf and floats in mid-air, remaining approximately the same height as the shelf it formerly inhabited. The man doesn't seem to notice as the box of tea (so we're told; we can't quite make out the box's label on the video) remains suspended in air. Then another box, yellow in color, launches itself from a shelf in front of the man and to his right, immediately dropping to the floor. This catches his attention. When he turns to look, the red box also drops.

Objects moving of their own accord in this manner are, in the paranormal lore, generally described as "poltergeist" phenomena. The word poltergeist comes from the German for "noisy ghost" and that's pretty descriptive. These are the paranormal phenomena that involve disruptive physical actions including loud noises and objects moving. Traditionally these are thought to be either a type of ghost or a specific behavioral classification for ghosts, though in contemporary paranormal lore, some poltergeist phenomena are thought to be caused by a variety of other supernatural entities including demons, psychics, elemental spirits, and more. Ghosts or spirits are probably still blamed for the majority of alleged poltergeist phenomena, with interpretations of their motivation ranging from malice to mischief to confusion.

Media representations of alleged poltergeist activity are often some of the more exciting ones. Plenty of ghost photos or videos just contain "orbs" (see Volume 1, Chapter 24), streaks of light, double exposures, or any number of other photographic anomalies that, even if one is inclined to supply a paranormal explanation, don't exactly get the heart rate going. Poltergeist videos are something else. We understand what we're seeing (even if we don't understand the cause), and so we have something to sort of hang our conceptual hats on when evaluating them. Plus, the idea that these supernatural entities can manipulate physical

---

1   Unless we were to turn this volume into a flipbook, which is an amusing thought but one we quickly abandoned once we sobered up.

2   Foxnews24x7 (July 25, 2013). The ghost of Earl Grey Teabags float down aisles of shop. [online video]. *YouTube*. <https://www.youtube.com/watch?v=QutV0mzLpms> (accessed June 23, 2024).

objects raises a number of questions and possibilities including the frightening possibilities of horror movies as well as the more intriguing possibilities of some form of communication with the "other side."[3]

They also tend to make us suspect hoaxes. Not to say all poltergeist videos or allegations necessary *are* hoaxes. In fact, we know some of them are not. The entire range of possibilities still exists in general terms, from genuine supernatural activity to psychological explanations to mistaken natural phenomena, and all the other things we always consider on one of our investigations. But the clarity of many poltergeist videos—including the tea ghost—seems to render certain explanations less likely. There's no way, for instance, this video could be psychological in nature because, after all, it was captured on video and seen by millions of perfectly sane individuals. We also can't think of any natural phenomenon that would possibly explain this kind of behavior from a tea box. Sure, we'll always leave open the possibility that there's some third explanation we just didn't think of, but we were pretty sure we were going to be dealing in this case either with a genuine supernatural event or a deliberate hoax.

Earlier in this book (and plenty of places elsewhere) we've mentioned that hoaxes are relatively rare. And that's true. People often ask us if we get a lot of hoaxes in our caseload and the answer is: not really. Of the cases we're actually specifically engaged to investigate by a homeowner or business owner, we can count on one hand the number of hoaxes we've dealt with (probably with some digits left over). Maybe it's because our philosophically neutral approach doesn't really invite people to try to prove us wrong with a hoax. Maybe it's because we have a reputation for being skilled at detecting the hoaxes we do find. Or maybe people just aren't that inclined to make these kinds of stories up. We're honestly not sure, but we're glad to say we don't have to spend too much time busting fakers.

Reports in the media are a bit of a different matter, though. In that arena, we see a lot more deliberate hoaxes. They're still, as far as we can tell, a severe minority of paranormal claims. But they are out there, and because the more spectacular videos are simultaneously more likely to go viral and more likely to be a hoax, we always keep our metaphorical antennae up for stories that could benefit from our skeptical analysis. The tea ghost ended up being such a case.

Chapter 6 already covered the idea that magicians' perspectives are often useful on a paranormal investigation and mentioned that several of our members have professional training as magicians, so we don't need to revisit that entire conversation except to remind you that these are skills we possess and modes of thought with which we're comfortable.

We'd also like to raise an issue briefly here which we'll tackle in a bit more detail in Chapter 24, and that's the idea that hoaxes, while always a hypothetical possibility, are sometimes ruled out as potential explanations by other investigators because the people commenting on a case don't believe a hoax could have been perpetrated with a small budget. Hollywood resources can do almost anything, the thinking goes, but amateurs working with simple "around the house"

---

3   Or, as one Internet meme put it: if poltergeists can slam doors and rearrange furniture, they can pick up a mop and contribute to the household.

materials ought not to be able to create something that looks so good. It's an argument we heard about the tea ghost video, and one of the things we wanted to determine was whether, using our skills, we could produce an equivalent video on a shoestring budget.

If you noticed the image on the title page for this chapter, taken from our own recreation video, you already know the answer was "yes." In fact, putting our magicians' training to use along with about five dollars worth of materials we already had lying around the house, we were able to create several variations of the floating tea box effect. We'd like to say we're exaggerating about the five dollars in materials part, but our solution was actually just that simple.

We're conflicted, here, about whether we should reveal the actual method we used. On the one hand, our magicians are reluctant to reveal any of the tricks of their trade. It's not so much a moral or ethical issue for us (as it seems to be for some magicians) but more of an aesthetic or artistic one. Magic tricks have simple solutions most of the time, and once you learn them, that sense of wonder we're always trying to create is gone forever. Killing that sense of wonder for someone is the last thing we want to do. But on the other hand, we also want to prove our case regarding how simple it was to recreate the floating tea box. Our final solution is this: we're going to put it into a footnote. If you really want to know the stupid-simple method we used and kill that sense of wonder, just read the footnote attached to the end of this sentence; but if you want to keep the magic alive, just skip this footnote and know that we're being honest when we say our method was cheap and pretty easy.[4]

Just to be thorough, we also wanted to see if we could additionally recreate the video using CGI special effects. None of us at the time were professional filmmakers (we've worked with filmmakers on some other projects in the years since and have even started toying with the idea of some film projects of our own, both individually and as part of our Rocky Mountain Paranormal activities), but we were familiar with the basics of the technology and wanted to know if we could make a convincing floating tea CGI video even as amateurs using off the rack video editing software. At the time, that was slightly beyond what we could do "in house," so we contacted a filmmaker of our acquaintance to see what it would take and if, in partnership with this artist, we could use CGI without Hol-

---

4   The secret method: thread. We used black cotton thread we already had lying around in one of our homes as part of a sewing kit. People initially discarded the thread idea because the floating boxes seemed to move too smoothly and steadily and didn't flop around while hovering in air as they would if hanging from a string. But we didn't just dangle it from a piece of thread. Instead, we used two long lines of thread to create a "rail" system to stabilize the box, and another piece to pull it off the shelf and onto our thread rails. To make it even easier, we carefully emptied the boxes and then resealed them so we didn't have to fight the weight of a full box of tea bags. Then we just adjusted the lighting until the threads couldn't be seen by the camera. We warned you it was going to be disappointingly simple. And we don't feel too bad about revealing a magicians' secret because professional magicians can make things float under entirely different circumstances with entirely different methods. This particular method is relatively limited to this kind of application.

lywood-level resources.

Indeed we could and indeed we did. We should clarify that the software we used would have cost more than the five dollars we mentioned. Video editing software has become cheap enough that it's readily available to amateurs and often comes as part of a software package. So while we can confirm that even the CGI part of our experiment would not have been cost prohibitive to most people, someone starting from scratch would have had to invest perhaps a few hundred dollars. At this point, a decade later, we'd be able to do it ourselves using video editing tools already included in software packages we use for our other business.

All told, by the time we were done, we had tea boxes sliding across tables, jumping off tables, hovering gently in mid air, flying across the room, and doing all manner of things tea boxes ordinarily ought not to do. Because video and print don't easily mix, you can find the entire recreation video online at the link in the footnote.[5] Our senses of humor have not matured in the decade since, so if we were to do it today, with the even more powerful and even easier to use video editing software now available to us, we'd probably end up with tea boxes singing and dancing by the end.

Here's an important point. Our experiment proved that videos remarkably similar to the one that made headlines could be achieved using natural and inexpensive means by a reasonably creative hoaxer. That's a pretty convincing piece of evidence but it is *not* proof that the original video actually was a hoax. Just because we can do something by trickery doesn't mean trickery is necessarily the only way to do it. Occam's razor suggests that we should be more inclined to believe trickery is the explanation unless there's some compelling reason to think otherwise, but an inclination in one direction or the other is not the same as proof.

The proof came a few months later when the hoaxers came forward with the truth. It turns out the whole thing was cooked up by Scottish magicians Barry Jones and Stuart MacLeod (who perform as Barry and Stuart) as part of a TV program they were producing called *The Happenings* in which they attempted to convince residents of a small town that something supernatural was afoot.[6] This revelation required Barry and Stuart's own admission on their TV show because the owner of the market involved had signed a non-disclosure agreement.[7] Barry and Stuart did not reveal the method they used to create the trick, only specifying

5   Rocky Mountain Paranormal (2013). Tea stealing ghost. [online video]. *YouTube* [channel: Warningradio]. <https://www.youtube.com/watch?v=UnhLvZ9gz6k> (accessed June 23, 2024).

6   Jones, B. & MacLeod, S. (Executive Producers) & Haug, J. (director). (December 17, 2013). Haunted Town (Season 1, Episode 3) [TV series episode]. In Jones, B., MacLeod, S., Bracken, D. Poesigger, J., & Cornwell, P. (producers). *The Happenings*. Objective Productions & UKTV Watch.

7   Kent Online (2013). Tea bag poltergeist at Whitstable Nutrition Centre exposed as fake by TV magicians Barry Jones and Stuart MacLed [sic] for The Happenings. *Kent Online*. <https://www.kentonline.co.uk/canterbury/news/poltergeist-exposed-10611/> (accessed June 23, 2024).

that it was a magic trick. It's entirely possible they could have used a different method than we did, especially since they did have a TV production budget backing their work. But we assume they probably came up with a method similar to ours.

Aside from just having fun with our recreation videos—and they were indeed fun—we were pleased to be able to show how easy it is even for low-to-no budget amateur operations to create the kind of hoax that could, at least hypothetically, become a viral sensation with millions of viewers. Our recreation video never got the same attention as the original in this case, though. For our own viral sensation, you'll have to read the next chapter.

24
The Alien in the Window

This and the following chapter are part of the same saga, but are separated for ease of understanding. Warning in advance: these two cases are real rabbit holes. If you like a good extraterrestrial conspiracy, we've got you covered!

Several years ago, a man named Stan Romanek recorded a video—now commonly referred to as the "Boo Video"—allegedly depicting an alien looking in his window. Few noticed. But in 2008, he appeared on the Larry King program alongside Jeff Peckman to endorse a political campaign centered on extraterrestrials (which will be the subject of Chapter 25) in which he cited his video as a key piece of evidence. The video went viral and Rocky Mountain Paranormal got involved, initially just planning to see if we could recreate the video, but then we saw how deep the rabbit hole actually went.

Figure 24.1. Rocky Mountain Paranormal's own alien in the window. Photo: RMP archives.

It's difficult to know what to include in this chapter, whose subject is the alien in the window video, and what to defer until our next chapter on the Denver Extraterrestrial Affairs Commission. These two cases are so intertwined that there's necessarily going to be some overlap. For ease of understanding, we've decided to present the alien in the window story, our recreation thereof, and our research into Stan Romanek, creator of the original alien video, in this chapter, and to put everything else in the next one.

Romanek's interest in UFOs and alien abductions has been long in the making. He's been seeing UFOs for decades, with a list of encounters including more than 195 "documented events" (his term, not ours).[1] Our analysis of many of his claims simply comes down to "mistaken natural phenomena" and isn't honestly all that interesting. Many UFO[2] photographs, videos, or even in-person encounters can easily be chalked up to such errors in discernment. Recall our Lafayette UFO case (Volume 1, Chapter 26) which turned out to be firefighting aircraft. Others have famously been weather balloons, atmospheric phenomena, a wide variety of aircraft, children's remote control airplanes, drones, deliberate hoaxes, and so much more.

For our purposes, it doesn't much matter whether Romanek personally believed in the extraterrestrial origins of these early encounters. For that matter, it doesn't even really matter whether any of them even took place. The action of the story comes later. What matters here is that Romanek had a long-standing interest in this subject and a knack for telling his stories in the kind of way that people were inclined at least to listen to him, if not to believe everything he said.

The part of his story that catapulted him into the limelight thanks to an interview with famed anchor Larry King in 2008 was his "Boo Video" showing an alien looking in his window. It was the classic image of a gray alien. Large hairless head, big eyes, small mouth, diminutive stature. Classic alien. And yeah, it was peeping in the window.

We initially caught wind that such a video existed a few years before a clip from the footage went viral after Romanek's Larry King appearance, but we hadn't managed to get our hands on the footage itself. Whispers were going about in the UFO community that this was going to be "the" video—the one that would finally settle the matter and bring the whole coverup to an end. Some people we knew in the local UFO community claimed to have seen it, but none of them could provide us with a copy. It was, they said, being kept under wraps pending evaluation by scientists.[3]

Eventually, though we couldn't get our hands on an actual copy of the vid-

---

1   Sumple, J. (Writer & Director). (2013). *Extraordinary: The Stan Romanek Story* [Film]. j3FILMS.

2   Many now prefer the term UAP for "Unidentified Aerial Phenomenon" rather than the classic UFO for "Unidentified Flying Object." It doesn't much matter to us because the operative word in both cases is "unidentified," but we're old school and stick to the classics.

3   In case there's any doubt, Larry King was not a scientist and his television program was not a scientific journal. But we digress.

eos (there were actually a couple of them, of which the window shot became the most famous), we did get a glimpse of them at a lecture we attended in which Romanek presented some of his evidence. Most of it was laughable, but the two videos were of slightly higher quality than most of the rest of what he brought.

In the first video, Romanek is awakened by a red dot on his ceiling which we believe to have originated from a laser pointer. When he got up to look around his house—for some reason bringing his video camera with him—he was flabbergasted to discover a small (roughly three to four foot tall) alien peeking at him around a hallway corner. Eventually even Romanek said he was going to stop exhibiting that video because it looked too fake. But his second video, the one of the alien looking in the window, caught on. In this second video, an alien's head pops into view, apparently peeping into Romanek's home window, looks around for a bit, blinks[4], then ducks out of view.[5]

Several claims were made by Romanek and his supporters to support the film's authenticity. For example, one "expert" said a hoax wasn't likely because the alien looked into a second story window, meaning if someone did create a hoax, they would have had to erect scaffolding to get the alien in place. Most significantly, it was suggested by some believers that it was unlikely to be faked because a hoax video of that quality would require at least a $50,000 special effects budget to pull off.

We thought otherwise. And when the video started getting national (and even some international) attention, we decided we needed to see what it would actually take for our amateur crew to pull off such a fake. The hardest part—and it wasn't very hard—was to find a good-looking latex rubber alien. We found one for sale for about $90 and placed our order. Once our little alien guy arrived, we knew we were in business. For such a low price, we'd purchased a prop that not only—we thought—rivaled what we'd seen in Romanek's video, but has now found a permanent home in our archives.

We set up our little amateur production using our prop alien and nothing else except a few tools and materials we happened to have in our homes already. First, we recreated the video as accurately as we thought possible, resulting in a video you can see online and from which we've printed a screen capture on the title page of this chapter.[6] And yes, ours also blinks. In fact, we think our blinking effect, supplied at no particular difficulty by our previously-mentioned CGI effects consultant, looks far more convincing than in the original. For the record, though, only the blinking eyes was done with computer animation. Everything

---

4   The blinking looked to us more like just a trick of the light, but that was part of the claim made about the alien, so we report it for what it's worth.
5   The full video, including Romanek's overacting, can be viewed in the documentary about his case: Sumple, J. (Writer & Director). (2013). *Extraordinary: The Stan Romanek Story* [Film]. j3FILMS.
6   The "final product" of our recreation is, as of this writing, available online at https://www.youtube.com/watch?v=1ema9Stirsc. For those interested in behind the scenes footage or just want proof that we recreated the video by the means we've described, a humorous outtake is also currently available at https://www.youtube.com/watch?v=wEFtnsqvbEQ.

else was accomplished practically with our prop alien. And none of it cost us anything close to the $50,000 effects budget such a video was supposed to require.

With regard to scaffolding being required for the alien to have reached Romanek's second floor window, there's an equally compelling and equally simple solution: we put the alien on a stick.

Figure 24.2. The alien reaches the second-floor window. Photo: RMP archives.

Though we felt we had done a pretty good job of showing that Romanek's video could easily have been accomplished by mundane means, we also wanted to dig into the backstory a little bit. And the claims get a lot stranger than merely witnessing an alien in the window.

Despite claiming only a limited mathematics education, Romanek published a series of equations he said had been revealed to him under hypnosis. While a complete treatment of the subject of hypnosis would require at least a chapter (if not a whole book) of it's own, let's just say we're skeptical. But we looked at the equations themselves and found them to be a mishmash of previously-published equations rather than anything that would require alien knowledge to have produced.

One of his equations looks like a standard differential equation, butted up against something about bombarding a helium atom with element 115. Understanding element 115 is a whole rabbit hole in and of itself. Long hypothesized based on a gap in the periodic table of the elements, the so-called element 115 gained attachment to paranormal lore when UFO enthusiast Bob Lazar claimed to have studied an extraterrestrial vehicle fueled by this element while working at Area 51. We could write a whole other book on Lazar's full story. As it turns out, ghost stories are relatively straight forward (not necessarily the same thing as being easy) to investigate. These alien stories end up being these massively cross-referenced collections of interrelated claims and conspiracies. Anyway, at the time of Lazar's pronouncement and Romanek's equations, element 115 hadn't yet been synthesized. It since has been (in 2003) and was given the name moscovium in 2016.

Another part of Romanek's mathematics includes the famous Drake equation, written by astrophysicist and astrobiologist Frank Drake as a kind of tool to organize one's thoughts about how to estimate the number of potentially intelligent civilizations that might exist in our galaxy. However, Romanek multiplied the right side of Drake's equation by a factor of 100 without explanation and apparently without any reason other than to inflate the estimate, calling his mathematical prowess into serious question.

For those unfamiliar with Drake's equation, it goes like this:

$$N = R_* \cdot f_p \cdot n_e \cdot f_l \cdot f_i \cdot f_c \cdot L,$$

where N is the number of potentially communicative civilizations in our Milky Way Galaxy, R_* is the average rate of star formation in the galaxy, f_p is the fraction of those stars with planets, n_e is the average number of "Earthlike" planets that could potentially support life per star with planets, f_l is the fraction of such planets that eventually develop life, f_i is the fraction of those life-bearing planets that eventually develop intelligent life, f_c is the fraction of intelligent civilizations that eventually develop communications technology capable of being detected by other civilizations (such as our own), and L is the amount of time such communicative civilizations release those signals into space.[7]

The Drake Equation isn't the kind of equation we can really solve. In fact, we can't do a whole lot with it because the vast majority of its terms are unknown to us. It's a valid mathematical expression in the sense that it would give a viable estimate of the number of potentially communicative civilizations in the galaxy, but only if we could supply correct numbers for each of its terms. Often we like playing around with it in our spare time, raising or lowering different numbers based on new astronomical observations or different sets of assumptions. But it's meant as a way to organize our thoughts, not as the kind of thing that actually gives us a final answer in the form of a simple number.

What bothers us in particular about Romanek's multiplication of the Drake Equation by 100 is that it seems entirely unnecessary. If one wanted to increase one's estimate of the number of communicative civilizations, one could just change one's estimate of any of the other terms in the equation. But if we assume, as Romanek seems to suggest, that the equation is being supplied to him by an extraterrestrial intelligence who presumably *knows* the correct values of each term in the equation (they'd almost have to, at least within reasonable margins of error, if they were technologically advanced enough to visit Earth), then the

---

7   SETI Institute (n.d.). Drake Equation. *SETI Institute*. <https://www.seti.org/drake-equation-index> (accessed June 24, 2024).

equation should give the correct number *as written*. Multiplication by 100 (or any other value) would render an incorrect value.

A running theme in Romanek's equations is the blending of disparate bodies of knowledge without rhyme or reason. In one place he'd write a meaningful equation but out of its proper context, and then he'd mush it up against Aramaic characters for no clear purpose.[8]

Our math skills are pretty good and we think we have above average understanding of astrophysics, but just to make sure we weren't missing something significant, we wrote to Dr. Seth Shostak, Senior Astronomer at the SETI Institute[9] for an expert opinion. His reply: "In fact, I did look at these equations. He could have cribbed them from just about anywhere...college texts, the web... even Wikipedia. They're no more proof that he's been in contact with aliens than his stories are. He has to show something NEW [emphasis in original]. Some fact we don't know. Not just rummage through the human attic and find a few things of which we're already aware."

Another extraterrestrial claim Romanek made not directly related to the alien in the window (but tangentially related as a part of his overall story of extraterrestrial experiences) is that one of his alien encounters resulted in some kind of mysterious device being implanted into his body. This supposedly occurred during one of several alien abductions of which he claims to have been victim. When ABC News asked him to submit the device for scientific or medical examination, he claimed it vanished from his body.[10] Photographs of his wounds glowing in mysterious colors under an ultraviolet "black light" can easily be reproduced by applying to the skin any number of substances that luminesce under such a light.

In a truly stunning display of what we can only describe as pure narcissism, Romanek also reported having received a voicemail message from a character called "Audrey" who spoke with a British accent (apparently through a voice synthesizer) which said, in part, "As you have probably noticed, Stan [Romanek] is slightly different. The way he thinks, the way he perceives the world, seems to be a little more advanced than usual. The interesting thing is that Stan has no idea who he really is." In another message, this time directly addressing Romanek, the

---

8   A large sampling of the equations and doodles he scribbled, apparently under hypnosis, can be seen (albeit without much further explanation) in the documentary made about his encounters: Sumple, J. (Writer & Director). (2013). *Extraordinary: The Stan Romanek Story* [Film]. j3FILMS.

9   SETI, or the Search for ExtraTerrestrial Intelligence, is an organization of legitimate scientists who have an interest in these questions. This is important because it proves that we're not accusing everyone who thinks about aliens of being a crank or a conspiracy theorist. These people are legitimate scientists who have dedicated their careers to working on this very question. The only difference is that SETI's experts, staff, and volunteers insist upon a high standard of scientific evidence that some other segments of the UFO-interested community don't.

10   MacGuill, D. (2017). UFO Enthusiast Now Facing Child Pornography Charges Is Subject of Netflix Documentary. *Snopes*. <https://www.snopes.com/news/2017/07/11/stan-romanek/> (accessed June 23, 2024).

voice referred to him as "Starseed."[11] The purpose in mentioning this piece of the puzzle is not to spend a lot of time analyzing the audio. It could have been produced without expense by anyone with a computer and its provenance is unknown. Rather, we simply think it's illustrative of Romanek's character and the scope of the story he was trying to tell through his various alleged encounters.

Lest you think we're being too harsh on Romanek, there are several additional points against the authenticity of his story. First, after his video gained national attention in 2008, he appeared on the *Coast to Coast AM* radio program during which host George Noory challenged him to take a polygraph lie detector test. He promptly failed and then later accused Noory of rigging the test against him.[12] Though we haven't looked into polygraph testing in our own experimentation because it's outside the purview of the paranormal, we have read quite a bit on the subject. We're skeptical of their accuracy in many cases, but they do detect common physiological responses to lying. We're not going to say that a false positive is impossible, but at the same time, we also don't believe George Noory rigged the test.

In 2015, he confessed. Sort of. He had been caught in the act of faking a paranormal manifestation during a live interview and had to admit he'd faked those "moving objects," but would go on to attempt to explain the hoax away claiming that mysterious and malevolent government agents had coerced him into perpetrating the hoax to discredit himself.[13] Soon thereafter, facing even more serious charges, he would again suggest that government agents hacked his computer to plant false evidence against him.

Whatever one might think of the quality of Romanek's videos or the extraordinary aspects of his story, one question that came immediately to our minds was: even if we assume the videos to be genuine, why in the world would he have been recording in the first place? Witnessing an alien in the window is one thing. Having a camera already running to capture it on film is quite another. In the documentary about his life, Romanek explained. He was recording the window because he thought there might be a peeping tom trying to get a look at his stepdaughter and he wanted to catch the culprit on camera.[14] That explanation seems plausible enough (though it does leave a foul taste in our mouth in light of information about Romanek that was revealed some years later and which we'll get to shortly).

His story about the other video—the one in which the alien seems to be

11   Sumple, J. (Writer & Director). (2013). *Extraordinary: The Stan Romanek Story* [Film]. j3FILMS.

12   Scherstuhl, A. (2009). In Kansas City, celebrity UFO-filmer Stan Romanek finds an audience of believers—and one reporter. *The Pitch.* <https://www.thepitchkc.com/in-kansas-city-celebrity-ufofilmer-stan-romanek-finds-an-audience-of-believers-151-and-one-reporter/> (accessed June 23, 2024).

13   MacGuill, D. (2017). UFO Enthusiast Now Facing Child Pornography Charges Is Subject of Netflix Documentary. *Snopes.* <https://www.snopes.com/news/2017/07/11/stan-romanek/> (accessed June 23, 2024).

14   Sumple, J. (Writer & Director). (2013). *Extraordinary: The Stan Romanek Story* [Film]. j3FILMS.

peeking around the corner inside the house—is even more troublesome. Romanek's excuse for recording during that particular experience was that when he woke up, he saw a naked figure run through the house. He assumed it was his stepson and so he turned on the camera hoping to film the unclothed child "to blackmail him."[15] Though we are of the opinion this story is false and that Romanek was most likely deliberately perpetuating a hoax with his video, that excuse struck us as "off." As it turns out, it may have been something of a Freudian moment in Romanek's onscreen interview, because in 2014, Stan Romanek was arrested on child pornography charges.[16]

These books are meant to be fun. Serious about history and science, yes. But at the end of the day, we're just here to have fun while we learn some new things and talk about ghosts and aliens and monsters. We don't want to get into…this kind of stuff. Even writing about what Romanek did feels quite uncomfortable. Because it's part of the story, though, we will explain Romank's charges. We're just not going to linger on this part of the story for too long because we desperately want to forget about it and get back to our ghosts and our aliens.

The charges were filed against Romanek following a lengthy investigation by the Department of Homeland Security in which Romanek was found to be in possession of hundreds of images which, according to the Homeland Security report, "contained images of girls that spanned an age range of approximately 5 years old to approximately 12 years old…In each of these photos, the girls are posed in a seductive manner and are exposing their breasts and genitalia."[17]

True to form, Romanek not only claimed innocence, but insisted that the images found on his devices were planted by hackers as part of a government conspiracy to discredit, intimidate, or silence him about his work on UFOs.[18]

After refusing a plea deal, he entered a plea of not guilty in 2016.[19] In august of 2017, a jury returned a verdict of guilty on the charge of possessing child pornography but not guilty on the charge of distribution of the same.[20] A few months later, Judge Susan Blanco of the 8th Judicial District sentenced Romanek to serve two years in a halfway house program as well as ten years of supervised probation as a sex offender, a condition of which prohibited him from having contact with any minors without permission from the corrections program or using electronic devices without supervision.[21] In 2018, his probation was re-

15   *Ibid.*

16   Murdock, S. (2014). UFO Conspiracy Theorist Stan Romanek Says Child Porn Charges Are Gov. Conspiracy. *Huffpost.* <https://www.huffpost.com/entry/stan-romanek-child-pornog_n_4964459> (accessed June 24, 2024).

17   *Ibid.*

18   *Ibid.*

19   Rieck, D. (2016). Attorneys postpone Romanek trial. *Loveland Reporter-Herald.* <https://www.reporterherald.com/2016/08/03/attorneys-postpone-romanek-trial/> (accessed June 24, 2024).

20   Hindi, S. (2017). Update: Romanek sentencing hearing rescheduled. *The Coloradoan.* <https://www.coloradoan.com/story/news/2017/08/08/romanek-found-guilty/549901001/> (accessed June 24, 2024).

21   Hindi, S. (2017). UFO author Stanley Romanek sentenced to 2 years to halfway

voked and an arrest warrant issued after Romanek missed several required therapy sessions. He claimed this was due to illness, but Judge Blanco, observing that Romanek was still not taking accountability for his actions, resentenced him to an additional ten years of supervised probation in 2020.[22]

It's important to understand that just because Romanek is a sex offender, that doesn't necessarily make the rest of his claims false. However, this conviction and his subsequent behavior do speak to his character, so we consider it relevant in that sense. Combined with his confession of having hoaxed another video and the relatively low quality of his evidence, it helps to further convince us that his entire story is an elaborate fabrication.

Equally important is to realize that just because we're drawing a conclusion about this particular individual and his evidence, that doesn't mean we're making the same accusations of all UFO witnesses or even claimed alien abductees. Hoaxes seem to be more common in this world than on the ghost side of things, but our opinion remains that most claimed witnesses are fundamentally honest people. Some of them may be mistaking natural phenomena. Others might be manipulated by unscrupulous therapists or media into misremembering what they saw. Perhaps even some of them legitimately encountered extraterrestrial beings. Our claim is not and has never been that there's no such thing as an alien, nor even that there's no such thing as alien abduction. On the contrary, our only argument is that each case must be individually evaluated on its own merits. Like our friends at SETI, we're really hoping to some day find that critical piece of evidence. It's just that our evaluation of Romanek's case has reached negative conclusions.

We were really disappointed in the outcome of this case. Though we were unconvinced by the alien in the window video as soon as we saw it, we were genuinely hoping and expecting that this would be one of our more fun cases. Recreating the alien in the window film with our own charming little alien was a delight, and we would have been perfectly happy to leave it there. It's unfortunate that this case ended up having a dark ending. Don't let that completely turn you off, though. Our own alien is still with us and we still love him. We just figure the less said about Romanek the better. All the more reason we're thrilled that sometimes people play our own alien video in press reports, mistaking it for the original.

But our saga doesn't end here. We mentioned toward the beginning of this channel that much of the national press attention came in 2008 following Romanek's appearance on the Larry King program with a man named Jeff Peckman. They were using Romanek's alien video to try to generate public support for a ballot initiative to create something called the Denver Extraterrestrial Affairs Commission. It is to that subject we turn our attention in the next chapter.

---

house for child pornography possession. *The Coloradoan.* <https://www.coloradoan.com/story/news/2017/12/14/ufo-author-stanley-romanek-sentenced-2-years-halfway-house-child-pornography-possession/952799001/> (accessed June 24, 2024).

22   Swanson, S. (2020). Loveland UFO author, convicted sex offender sentenced to probation again. *The Coloradoan.* <https://www.coloradoan.com/story/news/2020/11/30/colorado-ufo-author-convicted-sex-offender-resentenced-probation/6465405002/> (accessed June 24, 2024).

25

The Denver Extraterrestrial
Affairs Commission

As a sort of addendum to our previous chapter, we now present the one and only time we got more actively involved in a political campaign. Using Stan Romanek's alien in the window video in an effort to gain popular support, a campaign emerged led by one Jeff Peckman to create an "Extraterrestrial Affairs Commission" for the city of Denver. The initiative appeared on the ballot in November of 2010. Because tax money and the reputation of the City were both potentially at stake, we felt we had to get involved so we registered a political campaign in opposition to the measure under the name Mission for Inhibiting Bureaucracy (MIB) and began an educational campaign about the issue. This case also includes the only time (to our knowledge) we've been accused in an official ethics complaint submitted to a government body of being covert government agents. That's an unusual feather in our collective cap.

Figure 25.1. Mission for Inhibiting Bureaucracy (MIB) logo. Image: RMP archives.

As we pointed out earlier in the book, we're not a political organization. Some of our members over the years have been politically-minded people (and of those, they've come from all across the political spectrum), while others have been largely apolitical. Never in our decades of history have we officially endorsed any candidate for political office, and we don't have any plans to start. Just as we always have to respect the religious beliefs of our clients when we visit their homes, so do we have to respect everyone's right to hold whatever political views they choose. It would be inappropriate for a research and educational organization like ours to get involved.

An important exception to that rule emerges, though, when a political campaign is specifically centered on a paranormal claim. When that happened in 2010, we felt the need to get involved both because tax money was potentially at stake and because part of our mission is public education. In light of the national press attention such an unusual ballot measure received, we thought it important for voters and spectators to get the whole story. Therefore, we formed an official political organization (legally separate from Rocky Mountain Paranormal but including the same membership) in opposition to the Initiative. We called ourselves the Mission for Inhibiting Bureaucracy, or MIB for short. Yeah, you can tell we took ourselves seriously. We even started wearing black suits with green ties, capitalizing on the stereotypical "MIB image." We created the official "opposition" website and set about making media appearances to comment on the matter. Because the ballot measure itself was so unusual, we found the press got very excited to hear what we (as well as the Initiative's proponents) had to say on the subject.

It's difficult for us to figure out how best to present the information in this chapter, since most of our activities here involved commenting on the ballot measure itself rather than our usual investigative or experimental work. As such, a good place to start would seem to be to actually read through the full text of the ballot measure interspersed with our own commentary. On the other hand, actually quoting the full text would end up wasting ink and paper because legal documents are often redundant. Thus, we begin by walking through the measure, quoting some of it directly and summarizing the rest. For ease of reading, we'll present the text of the measure itself[1] in *italics* when directly quoting a full paragraph, with our own commentary in standard typeface.

The measure begins with a section of operative definitions, as any good legislation must. Specifically, they define "extraterrestrial intelligent being" as creature "assumed to exist or originate from outside Earth." They also define an extraterrestrial vehicle as a "non-living method of transportation" of similar origin and exopolitics as the "study of the key individuals…and processes associated with extraterrestrial life."

---

1   As officially published to the voters:
Denver Elections Division (2010). "Ballot Initiative 300 Language." *City of Denver*. Archived online at <https://web.archive.org/web/20101121205125/https://www.denvergov.org/Portals/639/documents/Campaigns/2009-6-9_ET_%20Initiative_Ballot_Title_%20and_%20Text.pdf> (accessed June 28, 2024).

Here we didn't necessarily take issue with the definitions themselves for the most part. We didn't particularly like the use of the word "assumed" in paragraph 1 because our opinion was (and remains) that legislation should be based on fact rather than assumption. And we also questioned the restriction of the definition in paragraph 2 to "non-living" methods of transportation. After all, if we're going to go ahead and assume we're talking about extraterrestrial beings, who or what is to say they couldn't use biological modes of transportation? Most UFO lore suggests their vehicles are mechanical rather than biological, but even our own technology occasionally starts to blur those lines. We are, after all, now capable of intentionally engineering biological organisms. It's entirely possible that advanced aliens could do the same with enough sophistication to use organisms as a mode of interstellar travel.

Were this just someone's personal paper about aliens, we wouldn't necessarily be so persnickety about it all. But this was proposed to be written into the laws of the land. People might make fun of lawyers for speaking in "legalese," but it's important for language written in law to be incredibly precise. Lawyers and legal nerds can cite all kinds of cases in which important decisions turned on seemingly minor points of grammar. For one example, a recent $5 million lawsuit was decided on the basis of the simple lack of an Oxford comma.[2] So it's not mere pettiness that makes us question the language in these definitions.

The next section goes on to define the legislative intent. While this section is not operative law, it's the portion of the measure judges and lawyers (and the public) would look to for guidance as to the intentions of the legislation should there be any interpretive disputes. Therefore, though most people probably just think of this section as prefatory, we considered it one of the most important.

The first paragraph in this section argues that the presence of extraterrestrial beings has actually been *confirmed*, not merely hypothesized, by a variety of credible-sounding sources, bodies, or experts.

Hold the phone! This is where we start to have some more serious issues with the Initiative, because it asked the voters to agree, as a matter of law, that "credible evidence" supported claims of extraterrestrial visitation. By this point, you know we're not trying to argue there's no such thing as aliens, but if there has been this kind of confirmatory evidence, we've never seen it. Yes, there have been whistleblower reports from individuals with varying degrees of credibility, but none of them have risen to the level of confirming visitation by aliens. Official government documents, similarly, have never confirmed extraterrestrial activity. At least not any that have been declassified and published. There have been discussions of those issues in official proceedings, yes. But if there is a sort of "smoking gun" document out there somewhere that proves the whole thing, it's certainly still classified and buried so deep that we don't even know where one might try to look for it.

Once again, it's fine to believe that the government is hiding evidence of extraterrestrials. Skeptical though we may be, we're certainly willing to entertain the idea that such evidence would be covered up if it existed. But it's simply inappropriate to introduce these kinds of assumptions—particularly when presented

---

2   O'Connor v. Oakhurst Dairy, No. 16-1901 (1st Cir. 2017).

as facts—into the law.

The next paragraphs merit direct quotation for more thorough examination.

2. Evidence of extraterrestrial beings has been known by United States presidents since President Franklin Roosevelt, including President Ronald Reagan as disclosed in his presentation to the United Nations regarding extraterrestrial matters;

First of all, President Reagan did not disclose knowledge of extraterrestrial visitors. He did, in fact, mention a sort of alien force during his address to the United Nations General Assembly on September 21, 1987 in New York. But it's important to get the full context of his quotation. The relevant portion of his speech went as follows: "In our obsession with antagonisms of the moment, we often forget how much unites all the members of humanity. Perhaps we need some outside, universal threat to make us recognize this common bond. I occasionally think how quickly our differences worldwide would vanish if we were facing an alien threat from outside this world. And yet, I ask you, is not an alien force already among us?"[3]

Sounds intriguing, no? It certainly sounds like President Reagan accidentally let slip that he knew something about an alien force visiting Earth. And that quotation has been repeated by UFO enthusiasts for almost forty years at this point. But it's been taken out of context, because his very next words clarify his meaning: "What could be more alien to the universal aspirations of our peoples than war and the threat of war?"[4] The President was poetically referring to our conflict with the Soviet Union as "alien," not to extraterrestrial vehicles, reptilian shapeshifters, little green men, or any of the other things people have associated with his speech in the years since.

In that sense, this paragraph would have introduced a known and established falsehood into the law, so we objected on those grounds. Further, the idea that Presidents have all known about aliens has never been confirmed. Sure, it's possible that presidents-elect get briefed on extraterrestrials before they take office, but to assume that we not only have to assume the unproved existence of extraterrestrial visitors and the unproved existence of a government conspiracy to cover it up, but we also have to assume that every president since Roosevelt has managed to keep his mouth shut about it. We're pretty sure at least some of those presidents, if they knew anything, would have lacked the capacity to keep quiet. We could be wrong, but even if we are, we still objected to unproved assertions being inserted into the law.

3. Since at least 1947, the United States government has denied knowledge of this evidence, thus placing the public and local governments at potential risk and disadvantage;

The year 1947 is an allusion to the infamous "Roswell incident" in which an extraterrestrial spacecraft allegedly crashed in the desert near Roswell, New Mexico. In the decades since, countless pages (likely numbering in the millions at this point, though we freely admit that's just a guess because we have no good way to

---

3   Reagan, R. (1987). Address to the 42d Session of the United Nations General Assembly in New York, New York. *Ronald Reagan Presidential Library and Museum.* <https://www.reaganlibrary.gov/archives/speech/address-42d-session-united-nations-general-assembly-new-york-new-york> (accessed June 28, 2024).

4   *Ibid.*

track every publication on the subject) have been dedicated to the subject. Some believe the affair was a cover-up of a legitimate extraterrestrial encounter. Most reputable scholars—even those inclined to believe in aliens or UFOs in general—however, have come to the more parsimonious (and at this point, very well documented) conclusion that it was a crashed high-altitude surveillance balloon used as part of the then-classified Project Mogul to monitor Soviet nuclear tests.[5]

Putting aside the debunked Roswell story, the idea that the people and their local governments might have to take action in the face of a Federal cover-up merits some additional commentary. For one thing, the cover-up itself could not conceivably be a single-nation affair. To keep secret the kinds of information this Initiative alleges, a worldwide conspiracy would likely be necessary.

But from a legal perspective, this also raises an important Constitutional question within American law. We're not going to say the formation of an ET Affairs Commission itself is a violation of the Constitution. It may be misguided, but it's probably not illegal.[6] However, this paragraph does at least hint at a Constitutional issue. The Constitution expressly makes it unlawful for any of the states (and by extension, municipalities) to "enter into any Treaty, Alliance, or Confederation...[or] without the Consent of Congress...lay any Duty of Tonnage, keep Troops, or Ships of War in time of Peace, enter into any Agreement or Compact with another State, or with a foreign Power, or engage in War, unless actually invaded, or in such imminent Danger as will not admit of delay."[7] That foreign affairs are not within the Constitutional purview of the states or municipalities is further clarified by the granting of foreign policy powers, with the advice and consent of the Senate, to the President of the United States[8], by the Supremacy Clause[9], and by a variety of other clauses granting foreign policy powers exclusively to the Federal government, not to the states or municipalities.

Could it be said that the aforementioned paragraph of the Initiative violates these Constitutional doctrines? Perhaps not, because this paragraph only expresses legislative intent rather than operative law. But it does at least raise the question: if we assume extraterrestrials are visiting, and a municipality enacted such a law, would the interactions between the municipality's commission and the

---

5   Klass, P. J. (1997). *The Real Roswell Crashed-Saucer Coverup*. Amherst, NY: Prometheus Books.
Note: there are any number of high-quality sources to which we could direct readers on this topic. A full exploration of the Roswell incident with all the relevant citations, however, is beyond the scope of this chapter. We've chosen the above as one of what we consider the best introductory sources on the topic.

6   As a tangential aside that fully captures the spirit of our own Constitutional analysis of this issue, the late Supreme Court Justice Antonin Scalia often remarked that he wished he could have a special rubber stamp made so that he could stamp a wide variety of laws as "stupid but Constitutional." See: Senior, J. (2013). In Conversation: Antonin Scalia. *New York Magazine*. <https://nymag.com/news/features/antonin-scalia-2013-10/> (accessed June 28, 2024).

7   Constitution of the United States of America, Article I, Section 10.

8   Constitution of the United States of America, Article II, Section 2, Clause 2.

9   Constitution of the United States of America, Article IV, Clause 2.

aliens not constitute foreign policy? The Founders probably never thought about extraterrestrials, but aliens would certainly have to be considered a foreign power, would they not? To be clear, we don't think there would be any Constitutional issues with a municipality forming a committee to study or discuss these matters, but the very moment they attempted to actually *do* anything about it, specifically if such activities might involve diplomatic relations with the aliens as the measure goes on to suggest, Constitutional issues would become relevant.

Paragraph four suggests that the United States (Federal) Government has engaged in a massive effort to cover up the truth about extraterrestrial visitation and that alien technologies so suppressed might include clean energy sources and "other advanced technologies."

Many people in the UFO and conspiracy theory communities have spoken at length about a variety of advanced technologies they believe are being politically suppressed. Of course, they never really have any good evidence to support those claims. True, many technologies are regulated by the federal government, and depending on the particular regulations and one's own political philosophy, some of those could rise to the level of "suppression" in at least some cases. That's ordinary politics and well beyond the scope of this book. For us, the relevant part of this paragraph was simply that it was once again asking voters to endorse claims for which there is insufficient evidence. It's entirely possible some technologies are being suppressed by political institutions. We'd be surprised if that weren't the case, frankly! It's even hypothetically possible some such technologies could be extraterrestrial in origin, though color us skeptical. But just like in middle school math class, you need to show your work.

For the record, one of the most common "suppressed technology" claims involves so-called "free energy" by means of some kind of perpetual motion device. This, however, violates the known laws of physics (the conservation of mass-energy assures that energy cannot be created nor destroyed, and the second law of thermodynamics tells us that in any exchange of energy, some portion of that energy will be entropically lost as heat). If someone thinks they have a discovery that could completely upend our understanding of physics, we certainly want to know about it—all science is, after all, tentative—but we're going to need to read the equations and inspect a working prototype before we're willing to enshrine that kind of belief in law.

The fifth paragraph in this section is a long one with numerous sub-paragraphs. It argues that municipal officials have ignored the "need" to address the extraterrestrial question. Support for that need is provided in the form of arguments that Federal agencies have confirmed alien visitation, that government and military leaders have described personal experiences with the aliens, that a "credible non-profit" has collected over 10,000 eyewitness reports of extraterrestrial activity. Paragraph 7 also makes similar claims. Nevertheless, Paragraph 5 goes on to suggest that ridicule has been the standard response from government officials when asked about these matters.

We think not. As the kids on the Internet like to say, "pics or it didn't happen." In other words: show us your evidence. No one we're aware of has ever seen these documents. Unlike alleged official documents, we are, in fact, aware of many of the UFO reports from a wide variety of the kinds of individuals mentioned. Most of them are people who reported seeing something unknown, which is entirely believable. Plenty of us see unidentified things all the time. Only a minority have ascribed their observations to extraterrestrials. Of those, we're not necessarily saying they're lying or crazy (though surely some of them could be either or both of those things), but none have been able to supply the kind of evidence required to settle the question.

The non-profit it mentioned was probably in reference to the catalog of UFO sightings maintained by MUFON, the Mutual UFO Network.[10] For decades now, MUFON has maintained a database of alleged UFO sightings as well as an investigative program to look into as many of them as they can.[11] Though they're less skeptical than we at Rocky Mountain Paranormal, we actually have a lot of respect for what MUFON attempts to do in documenting and researching anomalous experiences.[12] However even MUFON's International Director (until 2009) cited the Romanek alien in the window case (see previous chapter and recall that that was considered key evidence in support of this Initiative) as the kind of unconvincing evidence whose mishandling led to his eventual resignation from MUFON.[13] Not exactly a ringing endorsement. It's also impossible to determine what proportion of those reports are valid. While writing this chapter, we began work on an unrelated UFO case and determined that particular report to be valid (at least in the sense that a real person reported what he actually saw or believed he saw—whether there's anything otherworldly about it remains to be seen as we're only just beginning the case as of this writing), but while researching our case we also found dozens of other reports ranging from the easily debunked to the unverifiable.

At the same time, we certainly agree with the measure in the sense that we don't think blind denial or ridicule are appropriate responses to UFO claimants. They should be treated with respect and their claims should be evaluated on a case-by-case basis. Where we differ with the proposal, primarily, is that we think private organizations like ourselves or MUFON, with input from reputable scientists or institutions as necessary, should take the lead in these research programs. Until the evidence is a lot better than it presently is, it would be inappropriate to expend tax money or public time and energy on such an endeavor.

What the authors of the Initiative are trying to do here is a logical fallacy

---

10   Alternatively, they might have meant the National UFO Reporting Center (NUFORC), which is independent of, but often collaborates with, MUFON and other related organizations.

11   As of this writing, access to the database at the MUFON website at www.mufon. com is a benefit of paid membership in their organization and not available to the public. Members of the public can, however, use their website to report their own UFO sightings or alien abduction experiences.

12   We have sometimes taken issue with inconsistency in their evidentiary standards, but even if they don't always live up to the ideal, they're at least pretty good at saying the right things about open-minded yet skeptical investigation.

13   Carrion, J. (2010). Goodbye Ufology, Hello Truth. *Follow the Magic Thread.* <https://followthemagicthread.blogspot.com/2010/04/goodbye-ufology-hello-truth. html> (accessed June 28, 2024).

known as an "argument from authority." In logic, arguments are supposed to stand on their own merits. Expertise has its place, but it is not acceptable to believe or disbelieve a claim simply on the basis of the claimant's credentials. Lots of well-credentialed and intelligent people make mistakes. And plenty of uncredentialed people hit upon correct arguments. We need to see the actual data.

Paragraph 6 argues that part of the need for this new legislation is that the government of Denver has no procedures or protocols in place for evaluating claims of extraterrestrial activity or providing aid or services to those witnesses who make such claims.

Here we actually agree, at least in part. There are no such established protocols. By and large, reports to government agencies alleging UFO sightings or alien abductions or any other sort of paranormal phenomenon are either ignored or are dealt with by whatever first responders happen to be available at the time. Given the prevalence not only of UFO claims but paranormal claims in general, it would be better if first responders were trained to more effectively work with paranormal claimants. Rocky Mountain Paranormal would certainly be happy if police or other first responders would give us a call to advise and consult when paranormal claims are a part of their investigations, and we wish governments were better prepared to professionally deal with the varieties of religious or paranormal beliefs or experiences they might encounter.

However, doing so does not require the formation of a new bureaucracy. Or even if we were to desire a centralized government commission for such purposes, it would need to be created with a lot more philosophical neutrality than the proposed one. Furthermore, this paragraph contradicts the measure's previous assertions that the Federal government does have protocols in place to investigate (and then cover up) these kinds of phenomena.

8. The creation of an Extraterrestrial Affairs Commission will help to ensure the health, safety, and cultural awareness of Denver residents and visitors and, ultimately, facilitate the most harmonious, peaceful, mutually respectful, and beneficial coexistence possible between extraterrestrial intelligent beings and human beings.

There's a logical fallacy known as "begging the question." That occurs when an argument assumes the truth of a conclusion instead of providing evidence to support it. True, this Initiative doesn't claim to be about proving that extraterrestrials exist, but we consider it a part of the larger "UFO question" and in that context, they really seem to be jumping the gun here. Look, we'd be more than happy to discover aliens were visiting us (assuming they're peaceful, at least; though even if not, we'd still be interested to learn about them). And if and when that occurs, of course we'll need our governments to enact procedures to interact with them correctly. We'd even go as far as to say it's worthwhile for scientists to consider what first contact might look like both as a philosophical exercise and just on the off chance it ever actually happens. We just don't think we need government getting involved at this stage.

The next block of paragraphs goes on to list the following specific purposes for the formation of the E.. T. Affairs Commision.

1. To obtain and provide the most accurate, complete, credible, and relevant information available to city government personnel and residents about extraterrestrial intelligent beings on Earth;

Providing good information to the people is always a worthwhile goal. But if information is classified at the Federal level, this could easily promote some intergovernmental conflict. Further, we don't see why a government body is necessary to obtain and disseminate this information, particularly in light of the proponents' contention that such information has already been made publicly available (even though no skeptical researchers seem to have ever been given a copy of any of these documents).

2. To assist residents and visitors in reporting sightings of, or interactions with, extraterrestrial intelligent beings or their vehicles, and refer them to the proper and most appropriate public or private service agencies;

Another good cause. But once again, this is what MUFON already does. They do it privately. And organizations like Rocky Mountain Paranormal, also quite private, are always available to help people. We wish there were more of us and that we had more resources, but we don't see the need for this to become a public service.

3. To inform people of the implications of encounters with extraterrestrial beings or their vehicles;

What we wanted to know here was: how could we inform people of the implications of encounters—how indeed could we even begin to determine the implications of these encounters—without any documented contact with these alleged beings?

4. To develop protocols for peaceful and diplomatic contact with extraterrestrial beings in the event of contact;

Here we again encounter the Constitutional issues we raised above. Peaceful contact with extraterrestrial beings, assuming the ETs are peaceful themselves, ought to be possible. We hope. But to establish diplomatic contact with any foreign power, whether Earthly or otherwise, is not within the purview of municipal government.

5. To fund the Extraterrestrial Affairs Commission entirely from grants, gifts and donations, without requiring any fiscal outlay from the city budget; and

6. To make expenditures from the Extraterrestrial Affairs Commission fund.

We applauded the idea of funding this privately so as to not put the taxpayers on the hook for these expenditures. But we had questions. First of all, if this Commission wouldn't require any money from the public coffers, why would it need to be a public entity in the first place? There's nothing in law that prevents a private organization from advising government officials about whatever subject they might wish. Lord knows we don't keep our mouths shut when we know something we think our public servants need to know about, so why could the Extraterrestrial Affairs Commission not, quite similarly, form as a private organization dedicated to informing public officials about their concerns?

We were also concerned that, despite their stated goal to fund entirely from grants and donations, public money might end up being a part of this process anyway. According to the City's finance division, this Initiative would have cost at least $22,800 in the first year, with annualized costs and additional costs undetermined due to the novelty of the Initiative and unanswered questions regarding matters such as the cost of training first responders or other officials to respond to extraterrestrial claims. Even if the Commission itself could be self-funded, it would require contact with public employees whose pay comes from the public treasury, and the Initiative did not provide for those hours to be reimbursed. Further, we were told by one City official that if such a Commission were created in law, the city might have to pay for its maintenance out of the general fund in the event donations were not forthcoming.

Such a small amount of money, while significant to individuals like us, is basically a rounding error for a city the size of Denver with (as of this writing in 2024) a total budget of $4.06 billion and a general fund of $1.74 billion.[14] No taxpayer would frankly even notice the cost of one more small bureaucracy in such a large city. For us, though, it was about the principle of the matter much more than it was about the actual price. After all, when one starts adding up the costs of many such small bureaucracies, that's how we end up with governments spending as much money as they do.

The next section described the Commission's proposed staffing. Putting it mildly, we had some questions about the types of individuals it was meant to employ.

One expert in witness testimony from people who have previously had top secret security clearance and had direct personal close encounters with extraterrestrial beings or their vehicles. Such expert must have written at least one book or three scholarly articles focusing on whistleblower testimonies and/or five years work experience with an organization specializing in whistleblower testimonies.

This expert probably does not exist. If we add the word "alleged" to "close encounters with extraterrestrial beings," then we might be able to find a legitimate expert in the field. However, there is also a risk here of breaching the confidentiality associated with the top secret security clearance (or any level of security clearance, for that matter[15]). Even if the expert him/herself does not have such a clearance, it is a crime to knowingly disclose classified information to anyone not authorized to possess that information. The Initiative does not specify what means should be employed to avoid those legal quagmires.

One expert with a Ph.D. in the "natural sciences" i.e. physics, astronomy, or biology, etc.
One expert with a Ph.D. in the "social sciences" i.e. sociology, political science, etc.

These actually seem reasonable. Those are fields of expertise which would be implicated in any discussion of extraterrestrial beings, so those kinds of experts ought to be represented on such a committee.[16] We would change the lan-

---

14   City and County of Denver (2024). Adopted 2024 Budget. *City and County of Denver.* <https://www.denvergov.org/files/assets/public/v/3/finance/documents/budget/2024/2024-final-budget-book_ada-compliant_2-27-24.pdf> (accessed June 28, 2024).

15   Security clearances are often misunderstood. Some (but not all) of our members have held security clearances at various times as a result of their day jobs. Possession of a security clearance does not mean you immediately have access to whatever information you want. It means only that you've passed the requisite background checks to be able to view classified information when your specific job requires you to access those documents.

16   In fact, "astrobiology" has emerged as a legitimate interdisciplinary science in which one can earn either a graduate or undergraduate degree from a variety of high-

guage a little bit to account for different types of doctoral or professional degrees (that is, a D.Phil. or a Psy.D. would be just as qualified as a Ph.D.) and wouldn't mind seeing language allowing someone with "equivalent work experience" in lieu of a degree, but otherwise we agree.

A medical doctor who is a published author with expertise on the UFO/Extraterrestrial topic

That's a pretty small community. We agree that a medical doctor would be a reasonable expert to include. And given the subject matter, UFO experts (if we define expertise in that field carefully enough) would also make sense. But to require the same individual to be a physician, author, and UFO expert all at once seems like this line has been tailored to unduly restrict the mayor's choice to a very specific pool of candidates which might include individuals acquainted with the Initiative's proponents.

One expert with at least five years experience in investigations of UFO/Extraterrestrial encounters as evidenced in their authoring of at least one book or three articles in scholarly journal.

The problem here is that "scholarly journal" is not defined. By mainstream scientific standards, there are no scholarly journals that publish UFO encounters. But if one loosens the definition to include journals that do publish such material, one needs to define what qualifies a journal as scientific. We're not going to say there's no legitimate science to be done in the field. Our own case files demonstrate that we're committed to scholarly research on paranormal questions. And we suppose these books would have qualified us to serve on the Commission. But while some kind of UFO expertise would make sense for such a commission, the qualifications need to be either more carefully crafted or left more open to the mayor's discretion.

One expert with a certificate or diploma in Exopolitics or who works with a UFO/Exopolitical organization with a non-profit 501(c)(3) status in the U.S. or its equivalent in other nations.

The piling on of several distinct qualifications—here a certificate and a resume that includes work with a non-profit—again looks suspiciously like the qualifications have been written with a specific individual (or at least a small community of likeminded individuals) in mind. Further, we're unaware of any accredited institution of higher education in the United States or elsewhere that offers a certificate or diploma in "exopolitics." And absent accreditation, what would stand to prevent literally anyone from just printing off a certificate or issuing a diploma to one's own friends?

One expert who has consulted at least one hundred people regarding their alleged close encounters with extraterrestrial intelligent beings and has a Ph.D. in psychology or another suitable social science discipline.

This once again seems needlessly restrictive. However, including a psychologist (or psychiatrist or some similar mental health professional—in this context,

---

ranking universities. The study of astrobiology involves studying potentially habitable planets (or planetary systems), the search for life outside Earth, and the study of the chemical origins of life. As such, it employs research and expertise from such disparate fields as astrophysics, astronomy, biology, chemistry, biochemistry, geology, and more.

the difference barely matters) with some expertise in working with individuals who have claimed extraterrestrial experiences does make sense. If this were being done legitimately, such an individual would also have experience working with individuals who have claimed experiences but whose testimony was subsequently determined to be the result of either a hoax or of mental illness. Not that all such claimants are lying or crazy, but any legitimate expert would have to acknowledge that at least some of them are and should have experience working with those individuals.

The measure then goes on to list powers and duties. Some of our thoughts have already been addressed above, so we'll omit most of this section. However, in addition to the objections we've already noted, two specific paragraphs in this section stood out to us as meriting additional comment.

2. To obtain and make available on the official City of Denver web site the objectives of the commission, and any progress or other reports;

3. To display in the most cost-effective manner on the City of Denver web site, and otherwise, the most credible evidence and witness testimony regarding the existence and activities of extraterrestrial intelligent beings on earth;

We didn't object to these paragraphs in principle. However, publication on government websites requires a certain degree of expertise in information technology. Either the Commission would need to obtain the correct certifications or to burden established City IT employees with the extra workload. Furthermore, government websites have certain requirements concerning the materials they may or may not publish, and the Initiative does not clarify the process of getting the necessary approval for their publications. So even though this paragraph makes sense in light of the overall Initiative, it needs to be much more specific in its language.

The remainder of the text either rehashes issues we've already discussed or merely supplies a sort of boilerplate legalese to finalize the proposed law. We therefore omit the balance in the interest of brevity.

Our only further objection to the provisions contained in these final sections was that some were too vague to be useful and seemed to grant the proposed Commission perhaps too much self-governing power or too much authority to define its own scope. It's appropriate and common for a bill to include language to the effect that the relevant governing body shall have power to enforce or implement the bill by appropriate legislation. For example, many Amendments to the Constitution contain a clause declaring "[The] Congress shall have power to enforce this article by appropriate legislation."[17] The trouble here is simply that because the entire measure is so ambiguous in its scope, source of funding, and structure, it leaves many unresolved questions regarding what is necessary for proper implementation. Plus, it gave the relevant legislative and executive bodies only thirty days to execute such legislation, which is an incredibly short window. Government, sometimes by design and sometimes simply because of its size, moves slowly.

---

17   Constitution of the United States of America, Amendments XIV, XV, XVIII (repealed), XIX, XXIII, XXIV, and XXVI.

The measure appeared on Denver's ballot in November, 2010 as Initiative 300. It was a citizen's initiative, meaning it was placed on the ballot by a petition of registered voters rather than by the legislature. Its lead proponent, Jeff Peckman, initially submitted some 7,000 signatures but enough were invalidated[18] that he fell about 1,000 short of the requirement and had to scramble to make up the difference.[19] Peckman then resubmitted 10,000 signatures. Of those, 4,211 (just in excess of the required 3,974) were deemed valid and the measure was thus submitted to the voters for their approval or disapproval.[20]

So that was the measure, and we think we've done a pretty good job of spelling out most of our objections during our running commentary. In addition to offering similar commentaries both on our own website and, to the extent time and space would allow, in media appearances and interviews, we published an open letter in opposition to the measure. On what became the "official" opposition campaign website[21] we published a six-item list of our objections. Excluding the expanded reasoning behind each of the items (because we've covered the explanations in our commentary above), the six items were:

1. Initiative 300 promotes a belief system that space aliens are visiting us, based solely on stories but no actual physical evidence. While everyone is entitled to his or her own beliefs, the Denver City government should not be used as a platform to promote a belief system.

2. The supporters of the initiative claim it will be funded only through "grants, gifts and donations," and so will not cost the Denver taxpayers anything. The City budget office has already estimated its first year costs at $22,800. The initiative provides no evidence that it will receive any funding at all, and if it does not, the taxpayers could be stuck with the tab.

3. Everyone has a right to their own beliefs, their own faith, and to free speech. But without hard evidence, stories of extraterrestrial visitations remain in the category of a belief system, not science, law, or civics. Implementing this initiative, so as to make this law, would establish this belief system as part of our government, to which all its citizens must then be bound.

4. The initiative's commission requires each applicant to "be a knowledgeable expert in some area related to extraterrestrial intelligent beings or their vehicles." There are no standard tests, licenses, or degrees in "knowledge of extrater-

---

18   That's common. Any petition will have some invalid signatures. Some might be illegible. Others might not be registered voters in the correct jurisdiction. Given its subject matter, we suspect this one also probably got some "troll" signatures. Despite our opposition to the measure, we don't endorse that kind of behavior.

19   Calhoun, P. (2009). Wake-Up Call: ET, phone home—but use good manners. *Westword*. Archived at: <https://web.archive.org/web/20091018215859/http://blogs. westword.com/latestword/2009/10/wake-up_call_et_phone_home_--.php> (accessed June 29, 2024).

20   Osher, C. N. (2010). Denver set for UFO vote. *The Denver Post*. <https://www. denverpost.com/2010/09/28/denver-set-for-ufo-vote/> (accessed June 29, 2024).

21   It was at denveretcommission.org. Because the campaign is now over, we no longer maintain that domain, but most of the materials have been archived at rockymountainparanormal.com.

restrials." Therefore it is not possible for the Mayor to conscientiously implement this initiative, as he or she will have no criteria upon which to evaluate each applicant's credentials.

5. Supporters of the initiative often claim it will allow disclosure of secret government information on extraterrestrials and their technology. The ballot initiative simply creates a commission to communicate its opinions on UFOs. It does not give the commission any new special powers to force the government to disclose any secret alien technologies, or disclose the existence of aliens.

6. Initiative 300 promotional materials consist mostly of celebrity stories of UFOs (including, of course, from Elvis), and claims of government conspiracies (again, with no real evidence). But Initiative 300 is not about determining whether or not any of these claims are true. It just asks you to believe them, and to believe that space aliens are visiting us. If the supporters really had proof of this, they would show it to you now to ensure the initiative's passage. But they only present stories about UFO sightings and rumors and conspiracies. Without such proof, Initiative 300 is all flash, no substance, and just a big waste of time, money, and effort.

On that last point, we need to clarify that we obviously don't claim thinking about or discussing extraterrestrials is a waste of time. Obviously we're more than happy to spend our time on this kind of thing, and we honestly think more people should (especially if they can do so from a sober scientific perspective). Rather, our claim is that spending *tax* money on this sort of thing is wasteful[22] and that *governments* have better things to do with their time. Many perfectly good and legitimate activities are proper when done privately but would be an abuse of public resources.

From that point, the bulk of the rest of our activities surrounding this case amounted to a press tour. And it went viral. Because it was such an unusual ballot measure (it was the only local ordinance to be considered by Denver voters, but it appeared alongside an assortment of statewide measures on a variety of more "typical" subjects such as taxation, abortion, and gambling) it achieved not only local but national press attention. Our telephones and inboxes blew up. Unfortunately, as fun as it was to go on that little press tour as part of our opposition campaign, most of those appearances amounted simply to repeating what we've already described and so wouldn't make for very interesting reading here in this book.

But we weren't quite done yet. In addition to commenting on the measure itself, we also wanted people to know where it came from and what kind of evidence was being used to support the measure.

Chapter 24 already addressed the most prominent evidence and the charac-

---

22   Astute readers may recall that we joked about how much tax money we spent investigating a ghost story in our previous book (Volume 1, Chapter 14). However, in that case, the government became involved because there was a legitimate scientific question of interest to the government tangentially involved with the ghost story. We were more than happy to pool resources and let our paranormal investigation tag along with a legitimate geological survey. It would have been unconscionable for us to use public resources *only* for paranormal purposes.

ter of Stan Romanek. But though he was intimately connected to the case, even appearing alongside the campaign's initiator Jeff Peckman on the Larry King program to present his video evidence and begin gathering public support for the measure, the other players were also interesting characters, especially Peckman himself.

Peckman has a storied history both in paranormal claims and in politics. As early as 1998, he ran for one of Colorado's seats in the United States Senate as a member of the Natural Law Party.[23] He received 4,101 votes, or a grand total of 0.31% of the votes cast in that election.[24] Then in 2003, he spearheaded another ballot initiative in Denver aimed at promoting peace by reducing stress.[25] That certainly seems like a worthy cause (we would like less stress in our lives for sure), but one questions whether it can be achieved legislatively. The voters rejected the measure.

Peckman's push to de-stress Denver was likely tied to his personal philosophy, and perhaps to one of his business ventures. He studied at the Maharishi University of Management in Iowa, was (and perhaps remains) a practitioner of transcendental meditation[26], and founder of something called "Metatron Technology" which claimed to be able to protect customers from harmful electromagnetic radiation by converting it into "desirable healthy energy" through use of a variety of gizmos and computer software.[27] We're not saying the ballot measure was intended specifically to drive Denver residents toward Peckman's business interests, but it was surely at least motivated by his passion for the subject.

To do our due diligence, we obtained some of the Metatron Software called

---

23   Natural law is a serious subject in legal scholarship. To some religious legal scholars, it reflects an idea that, superior to the laws of man, there are certain God-given legal principles which must be respected. To secular scholars, natural law reflects principles, again often thought superior to statutory law, derived from observation of human nature and application of reason. However, the Natural Law Party (dissolved at the national level in 2004 and now active only in Michigan) was a small "third party" political organization dedicated to solving the nation's problems through transcendental meditation.

24   FEC (1998). 1998 U.S. Senate Results. *Federal Elections Commission.* <https://www.fec.gov/resources/cms-content/documents/FederalElections98_SenateResults.pdf> (accessed June 29, 2024).

25   American Morning (2003). Stress Test: Peace Ballot Initiative Approved in Denver. *CNN.* Transcript: <https://transcripts.cnn.com/show/ltm/date/2003-08-13/segment/06> (accessed June 29, 2024).

26   Transcendental Meditation (or TM) is a form of meditation developed by Maharishi Mahesh Yogi as a non-religious (though some scholars disagree with that categorization) way to reduce stress and attain higher levels of consciousness. It has attracted many celebrity practitioners but its claims to be any more effective than other forms of meditation are dubious.

27   Sutton, D. (2008). Alien ambassadors: A bizarre ufological scheme and an alleged ET video in Denver, Colorado. *Fortean Times.* Archived at: <https://web.archive.org/web/20140107181713/http://www.forteantimes.com/strangedays/ufofiles/1206/alien_ambassadors.html> (accessed June 29, 2024).

the "Metatron Global Peace Program" and examined it in detail. The installation notes said that the parameters would be specifically tailored to one's individual computer, but when we installed it on different computers we could detect no differences in the files or their contents. We also could not find any executable files included, which means it didn't seem to actually contain any programs a computer could run. It did, however, contain an approximately 3 MB encrypted PDF file which the notes claimed was essential to the program's proper functionality. When decrypted, we found the file actually contained nothing more than an image of a mandala.

Though extremely skeptical that such an image could have any effect on electromagnetism or a computer's functionality, we do have to remain agnostic when it comes to mystical or religious significance of such images (one individual suggested the image had been prayed over before it was encrypted). Despite the scientific-sounding claims associated with Metatron Technology, we have to recognize there is a religious element to this. Mandalas are devotional images in Buddhism, after all. Metatron itself, as a name, is a reference to the angel Metatron recognized by the Abrahamic faiths of Judaism, Christianity, and Islam. In some Rabbinic literature, Metatron is thought to be the celestial scribe, though he is not mentioned by name in the Torah or the Bible. So we had to keep our minds open to a potential religious significance to the file. However, we did point out that because the image was encrypted, any computer on which it was stored would not have access to the image itself prior to decryption. It would exist only as a mathematical expression, so we questioned its efficacy *even if* the image of a mandala might itself have some use.

We also found the general claims surrounding Metatron Technology to be dubious at best and a violation of the known laws of physics at worst. But since they appear (as far as we can tell) to be out of business now and not directly related to the ET Affairs Commission, a full discussion of their products is beyond the scope of this chapter. We mentioned what we did only as an important part of the chief proponent's backstory.

Speaking of which, Peckman wasn't finished in politics after the ET Affairs initiative. In 2011, he ran unsuccessfully for the office of Mayor of Denver. Out of 113,367 total votes cast, he won only 796.[28]

Beyond the two individuals we've mentioned, the campaign relied entirely on expertise offered by UFO enthusiasts rather than a balanced group of differing perspectives, as a legitimate government panel on the subject would have to do. Furthermore, many of the experts the campaign cited also worked on the Romanek case, and we've already seen how that turned out.

November came around and the voters finally had their opportunity to replace all the talking heads (ourselves admittedly included) as the final voice on the subject. Their voice was quite loud on the subject. When the *Denver Post* reported the election's outcome, incomplete tallies from the City Clerk's office had

---

28   Elections Division (2011). Unofficial final results. *City of Denver*. Archived at: <https://web.archive.org/web/20110507104618/http://www.denvergov.org/elections/CurrentElectionResults/tabid/430130/Default.aspx> (accessed June 29, 2024).

more than 84% of voters rejecting the measure.[29] In the final tally, that number dropped to a slightly (but barely) less embarrassing 82.34% who said no to the Extraterrestrial Affairs Commission.[30]

Initially, Peckman said the result was expected and that the campaign was more about starting a public conversation than actually changing the law.[31] However, he followed up his campaign by filing an ethics complaint against with the Denver Board of Ethics. Though ostensibly an ethics complaint against Denver financial staff for what he believed to be incorrect information about the Commission's potential costs, Peckman spent the majority of his 29-page document complaining about Rocky Mountain Paranormal (by name) and accusing our members of being "very disreputable, incompetent, unethical, and hostile persons" possibly secretly employed by "the CIA, NSA, or other covert group" to discredit his evidence.[32] The Board dismissed his case, citing lack of jurisdiction and lack of any complaint that could be adjudicated by the Board.[33] Just for the record, since the Board of Ethics never even got to the point of evaluating Peckman's bizarre claims, if we were actually employed by covert government agencies to hide the truth about extraterrestrials, we'd live in much larger houses, drive much more expensive cars, and wouldn't be writing books about the paranormal.

But since we're not getting that kind of secret money, we're going to have to keep looking into reports of alien activity with the same open minded but skeptical approach we apply to everything. We're hoping we eventually find the aliens. In fact, it seems quite likely that there are other life forms out there somewhere. Maybe (less likely, but still maybe) they're even visiting us. The evidence just isn't in yet to be able to say it with any degree of certainty.

29   Osher, C. N. & Glazier, K. (2010). Denver voters shoot down otherworldly measure. *The Denver Post.* <https://www.denverpost.com/2010/11/02/denver-voters-shoot-down-otherworldly-measure/> (accessed June 29, 2024).

30   Ballotpedia (n.d.). Denver Extraterrestrial Affairs Commission Creation Referendum (2010). *Ballotpedia.* <https://ballotpedia.org/Denver_Extraterrestrial_Affairs_Commission_Creation_Referendum_(2010)> (accessed June 29, 2024).

31   Osher, C. N. & Glazier, K. (2010). Denver voters shoot down otherworldly measure. *The Denver Post.* <https://www.denverpost.com/2010/11/02/denver-voters-shoot-down-otherworldly-measure/> (accessed June 29, 2024).

32   Peckman, J. (2011). Inquiry/Complaint Form. *City and County of Denver.* Archived at: <https://web.archive.org/web/20120303125501/http://www.extracampaign.org/uploads/INQUIRY-COMPLAINT_FORM2010_Peckman.pdf> (accessed June 29, 2024).

33   Denver Board of Ethics (2011). Digest of Selected Opinions January 1-June 30, 2011. *City and County of Denver.* Archived at: <https://web.archive.org/web/20140107173715/http://www.denvergov.org/Portals/5/documents/DigestOfOpinionsJan_June2011.PDF> (accessed June 29, 2024).

26

The Relic of Padre Pio

From aliens to religion, we cover it all. This final chapter concerns a Catholic Saint, Pio of Pietrelcina, more popularly known as Padre Pio. Numerous supernatural claims surrounded Padre Pio during his life, but our involvement in the Padre Pio story began when we acquired a remarkable old book we thought might have belonged to the Saint and so we set out to validate the relic and document its provenance.

Figure 26.1. The relic of Padre Pio (with third class relics). Photo: Bryan Bonner.

Catholics are known for their veneration of Saints. In fact, one of the major distinctions between Catholic and Protestant Christians has to do with this veneration. Protestants and non-Christians may find it odd, but Catholics are deeply dedicated to the collection, preservation, and veneration of their Saints' relics. For example, the head of Saint Catherine of Siena is preserved and displayed in an ornate reliquary at the Basilica of San Domenico in Siena, Tuscany, Italy. Or of local interest: the Cathedral Basilica of the Immaculate Conception in Denver is currently home of the tomb of Servant of God[1] Julia Greeley, a Denver-based former slave whose charity work led to her cause for canonization being opened by Archbishop Sameul Aquila in 2016.[2]

Several years ago, we—quite by happenstance—acquired an old Catholic book whose characteristics led us to believe it might have once belonged to Saint Padre Pio. Both because of the historic importance of the potential relic (y'all know by now how much we love history) and because of the religious and supernatural claims surrounding Padre Pio, we thought it important to learn as much as possible about the book's history and to validate whether or not it actually might be a genuine relic of St. Pio.

To begin with, we need to clarify just what a relic is in the Catholic tradition. To Catholics, relics are considered "sacramentals," a class of rituals and objects which do not confer grace in and of themselves (as they believe the seven sacraments of baptism, confirmation, eucharist, penance or confession, anointing of the sick or extreme unction, holy orders, and matrimony do) but whose practice or veneration can help point people toward holiness.[3] Until quite recently, official Church doctrine divided relics into three classes: first class relics consisted of any body part of a saint (or any relic of Christ himself such as a piece of the True Cross), second class relics included any item associated with the life of the saint such as a Bible or a piece of clothing owned by the saint, and third class relics were any item touched to a first class relic.[4] However, the Congregation for the Causes of Saints revised the classification in 2017 and relics are now considered either "significant" (whole bodies or significant body parts) or "non-significant" (fragments of bodies or what would have been considered second class relics).[5]

---

1   The title "Servant of God" is given to individuals whose cause for canonization as a saint is currently under investigation by the Church. It's the first step in the process. Later titles include Venerable, Blessed, and finally, Saint.

2   Julia Greeley Home (n.d.). On the Path to Sainthood. *Julia Greeley Home.* <https://juliagreeleyhome.org/on-the-path-to-sainthood/> (accessed July 1, 2024).

3   Catholic Answers (n.d.). Tract: Relics. *Catholic Answers.* <https://www.catholic.com/tract/relics> (accessed July 1, 2024).
Catechism of the Catholic Church (2nd ed.): paragraphs 1667-1679.

4   Mangan, C. (2003). Church Teaching on Relics. *Catholic Education Resource Center.* <https://www.catholiceducation.org/en/culture/catholic-contributions/church-teaching-on-relics.html> (accessed July 1, 2024).

5   Congregation for the Causes of Saints (2017). Instruction: Relics in the Church: Authenticity and Preservation. *The Holy See.* <https://www.vatican.va/roman_curia/congregations/csaints/documents/rc_con_csaints_doc_20171208_istruzione-reliquie_en.html> (accessed July 1, 2024).

Sale or incorrect disposal of "sacred relics" (meaning the first or second class relics addressed by the Congregation for the Causes of Saints' teaching) without authorization by the Apostolic See is forbidden by canon 1190 of the 1983 Code of Canon Law, but trade in third class relics (still popularly recognized by Catholics even though they're no longer considered official "sacred relics" by the Church) is common and permitted.[6] Most Catholics still think in terms of the three classes of relics. We'll discuss the classification of the potential Pio relic shortly when we get to our process of documenting its provenance.

Just as the previous chapter required us to exercise all due caution to constrict our political activity to the issue at hand and maintain our neutrality on other political matters, so does this case require the same level of care with regard to religious views. As such, we should start out by making it clear that Rocky Mountain Paranormal doesn't endorse any particular religious view. However, sometimes our work requires us to comment on a matter with some religious implications. Our strategy in such situations is to remain as neutral as possible while still being entirely truthful about the business at hand. In other words, if we were to discover some evidence that either supports or refutes a religious idea, we wouldn't be shy about publishing our findings (narrowly constricting them to only the views we either proved or disproved). However, absent that kind of evidence, we take religion as a matter of personal faith or philosophy. We're happy to describe religious views when they're relevant, but we don't ever want to come across as if we're evangelizing or criticizing.

We will have to take the same approach here. Padre Pio's life was marked by numerous allegations of supernatural occurrences. We will not be shy about describing those. We'll even address how the Church responded, both positively and negatively, to a variety of claims. But we were not present when these things were said to occur, so our ability to comment on them is limited. Our job in this chapter is to catalog the variety of claims made about Padre Pio and then to share the results of our investigation into the alleged relic, not to convince people to believe or disbelieve in the story of Padre Pio.

A good place to start would be to introduce readers to the life of the Saint in question. However, a full biographical treatment is beyond the scope of this chapter so we'll only hit some of the highlights of Pio's life before turning our attention to the supernatural claims and then to the relic.

Padre Pio was born to peasant farmers in Pietrelcina, Italy on May 25, 1887 and baptized Francesco Forgione two days later (it has later been pointed out that his name was significant as he became something of a "second St. Francis"). Even as a child, he was dedicated to Catholic life. His family attended daily Mass and he served as an altar boy at his local parish despite his poor health even from a young age. At the young age of twelve years, he expressed interest in becoming a friar, but required additional schooling. To pay for his education, his father took work in America and sent money home. By the time he was fifteen years old, young Francesco was ready for religious life and joined the Capuchin friars at Marcone, taking the religious name "Pio" in honor of Pope Pius I. He became a

---

6   Scripture Catholic (n.d.). Catholic Relics. *Scripture Catholic.* <https://www.scripturecatholic.com/catholic-relics> (accessed July 1, 2024).

priest in 1910, then was conscripted into military service during World War I in 1915, but was discharged due to his poor health in 1916. He spent the rest of his life serving in his capacity as a priest (which chapter in his life corresponds with the supernatural claims we'll address shortly). He died on September 23, 1968.[7]

During his service as a priest, Padre Pio was known for his religious zeal. When he heard confessions, long lines would form even though he had a reputation for being stricter than most priests, often withholding absolution until individuals corrected various elements of their lives; similarly, his Masses, though often exceeding four hours in length, were always well-attended and consistently drew pilgrims, some of whom called his celebration of the Mass "an authentic supernatural spectacle."[8]

On that note, Pio was the subject of numerous supernatural claims throughout his life. Some manifested only internally and therefore are the kinds of things that are beyond ordinary analysis, while others manifested externally and have been the subject of much discussion over the years. It is to those we now turn our attention.

According to the late Father Gabriele Amorth—himself famous as the exorcist for the Diocese of Rome (that is, Vatican City) and founder of the International Association of Exorcists—in his memoir about the Saint, Padre Pio's life was characterized not only by extraordinary holiness but by a constant battle with the forces of evil. Pio, said Amorth, regularly received visions and/or witnessed apparitions, beginning at about the age of four and extending through his entire life, "of his guardian angel, the Lord, the Blessed Mother, and others; the demons generally took the forms of wild animals, fearsome and threatening."[9] As a teenager, one of these visions informed Pio that his life had been chosen for a "great mission," and in another vision, he fought a giant which represented demonic forces. This was to be the beginning both of the Saint's holy life but also of a string of constant attacks from the forces of evil. For his part, Padre Pio welcomed the challenge. He considered his suffering to be meaningful. In 1910, he offered himself as a "victim soul" for sinners and the souls in purgatory. This is a concept in Catholic teaching that's considered "private revelation" and thus not binding on the Catholic faithful; in brief, a victim soul is thought to be a person whose redemptive suffering is joined to that of Christ.[10]

These are also the kinds of claims for which there's never really any kind of evidence. When one evaluates claims of the paranormal, one always looks for objective evidence or independent corroborating testimony. Visions or private apparitions, however, appear only to those who experience them. All one could do would be to examine the contents of the claimed visions to determine whether they comport with known facts, or to examine the neurological health of the

7   Catholic Online (n.d.). St. Padre Pio. *Catholic Online*. <https://www.catholic.org/saints/saint.php?saint_id=311> (accessed July 1, 2024).

8   Amorth, G. (2016, 2021). [Trans: Sherry, M.]. *Padre Pio: Stories and Memories of My Mentor and Friend*. San Francisco, CA: Ignatius Press.

9   *Ibid.*

10   Korson, G. (n.d.). What is a Victim Soul? *Simply Catholic*. <https://www.simplycatholic.com/what-is-a-victim-soul/> (accessed July 1, 2024).

experiencer, but one cannot directly know the mind or experiences of another.

External signs of Padre Pio's supernatural claims were not far behind, however. He was said to be a recipient of the stigmata, or the supernatural manifestation of wounds corresponding to those received by Christ during the crucifixion. These wounds, it is claimed, first appeared "invisibly" at the time when he accepted his status as a victim soul, but then became visible beginning September 20, 1918 and remained with Padre Pio through the rest of his life, only again disappearing, as suddenly as they appeared, upon his death in 1968.[11]

Importantly, though it's claimed that Padre Pio was attacked by demonic forces throughout his life, spiritually, psychologically, *and* physically, the stigmata are thought to be a blessed event and a sign of that aforementioned redemptive suffering. In Padre Pio's case, he was said to have gracefully accepted the wounds themselves but didn't like them to be seen publicly, so he often wore fingerless gloves to mask the wounds on his hands.

However, the word got out of a stigmatist in Italy and Padre Pio's popularity began to grow. Doctors and Church officials began to question the occurrence of the stigmata. Though Catholic doctrine admits of such a possibility, it's also well known that many reported cases of the stigmata have been the result of hoaxes or trickery. As is the case with any miraculous occurrence, the Church tries to balance their willingness (even eagerness) to believe in such things against a level of caution in the knowledge that an exposed hoax could damage their reputation. Though the faithful began flocking to his church to see what they now considered to be a man of extraordinary holiness, many also suspected Padre Pio could have been either perpetuating a deliberate hoax or suffering delusions.

One scholar suggested Pio may have faked the wounds by continually exposing himself to carbolic acid.[12] In support of that hypothesis are documents which confirm the Padre did indeed request such acid for a pharmacist, but believers are quick to point out that his stated purpose for wanting it was to disinfect needles for vaccinations and that there's no direct evidence he ever applied it to himself, nor even that the wounds necessarily were of the sort that could be caused by application of a concentrated acid.

Another hypothesis, suggested by Dr. Amico Bignami, was that the wounds initially appeared as a form of psychosomatic illness as a result of suggestion and then perpetuated by Padre Pio's use of iodine to disinfect them.[13] While a plausible idea at the time, as iodine was thought to potentially interfere with the regeneration of cells necessary to wound healing, more recent research has shown that using iodine as an antiseptic does not, in fact, impede, delay, or prolong healing.[14]

11   Amorth, G. (2016, 2021). [Trans: Sherry, M.]. *Padre Pio: Stories and Memories of My Mentor and Friend.* San Francisco, CA: Ignatius Press.

12   Moore, M. (2007). Italy's Padre Pio 'faked his stigmata with acid.' *The Telegraph.* <https://www.telegraph.co.uk/news/worldnews/1567216/Italys-Padre-Pio-faked-his-stigmata-with-acid.html> (accessed July 1, 2024).

13   Ruffin, C. B. (2018). *Padre Pio: The True Story* (3rd ed.). Huntington, IN: Our Sunday Visitor.

14   Vermeulen, H., Westerbos, S. J., & Ubbinkn, D. T. (2010). Benefit and harm of iodine in wound care: a systematic review. *Journal of Hospital Infection, 76*(3): 191-199.

Physician Agostino Gemelli thought Pio's wounds may have been self-inflicted, perhaps in a state of religious hysteria, and could have been preserved (if not originally caused) by continual reapplications of carbolic acid.[15]

On the other hand, numerous other doctors seemed a lot more convinced of the supernatural origin of Padre Pio's wounds. For example, Doctors Luigi Romanelli and Georgio Festi both examined Pio's wounds and found them scientifically inexplicable.[16]

At this point, there's unlikely to be any evidence that could finally decide the matter in either direction. Faithful will be inclined to believe in Padre Pio's stigmata and skeptics will be inclined to think they were either self-inflicted as a hoax or accidentally perpetuated by some environmental cause. Believers will point to the various physicians' inability to come to any consensus regarding a natural cause. Skeptics will point out that many of those physicians were associated with the Church and so may have had conscious or unconscious biases in their evaluations. Because Rocky Mountain Paranormal didn't have the opportunity to investigate Padre Pio's alleged stigmata for ourselves, we maintain neutrality on the subject and simply document the claims and various arguments so readers can form their own opinions.

Other supernatural occurrences about Padre Pio have been reported but without the same level of forensic study or validation. These include that in 1918, he experienced a state of altered consciousness known as a transverberation upon the appearance of his stigmata, and that he had the ability to bilocate, or to appear in two places at once.[17] One report even held that on a particular occasion Padre Pio managed to levitate his way to the confessional so as not to be seen and bothered while walking in; another story has him turning invisible for the same purpose.[18] A common claim made of the Padre was that he had the ability to read hearts or souls in confession and by that means determine whether someone's contrition was genuine prior to offering absolution (which, as we already mentioned, he would often withhold for the sake of the penitent's spiritual development). Father Amorth also claimed in his memoir that Pio could discern whether an individual was demonically possessed and had the ability to cast out demons.[19]

Of course these are the kinds of claims we cannot even begin to verify.

Bigliardi, P. L., Alsagoff, S. A. L., El-Kafrawi, H. Y., Pyon, J., Wa, C. T. C., & Villa, M. A. (2017). Povidone iodine in wound healing: A review of current concepts and practices. *International Journal of Surgery, 44:* 260-268.

15   Higgins, M. W. (2006). *Stalking the Holy: The Pursuit of Saint Making.* Toronto, ON, Canada: House of Anansi Press.

16   Amorth, G. (2016, 2021). [Trans: Sherry, M.]. *Padre Pio: Stories and Memories of My Mentor and Friend.* San Francisco, CA: Ignatius Press.

17   Castelli, F. (2011) [trans: Messori, V., Bockhorn, L., & Bockhorn, G.]. *Padre Pio Under Investigation: The Secret Vatican Files.* San Francisco, CA: Ignatius Press.

18   Plese, M. (2006). The Miracles of St. Padre Pio. *A Catholic Life.* <https://acatholiclife.blogspot.com/2006/02/miracles-of-st-padre-pio.html> (accessed July 1, 2024).

19   Amorth, G. (2016, 2021). [Trans: Sherry, M.]. *Padre Pio: Stories and Memories of My Mentor and Friend.* San Francisco, CA: Ignatius Press.

Many of them have been passed down for so long we don't even know the identities of the people who first made those claims. Even if we did know the exact origin, there's no way we could investigate. All we can really say is that, given the lack of evidence, some skepticism may be in order. Claims such as these are often added on to pre-existing stories as legends begin to develop, so it's entirely possible even if one believes in the fundamental elements of the Pio story (such as the stigmata) to imagine that the lore grew beyond the original facts of the case in the retelling.

A particularly interesting episode in Padre Pio's legend is that he prophesied the papacy of Pope John Paul II. According to the legend, Pio was visited by a young Polish priest named Father Karol Józef Wojtyla (who would later become Pope John Paul II). After hearing the his confession, Pio reportedly told the young priest that he would ascend to the highest post in the Catholic Church.[20] The Pope's secretary, Cardinal Stanislaw Dziwisz, would later deny this prophecy ever took place.[21] For his part, Father Amorth suggested only the individuals involved possessed the knowledge to resolve the debate (and they're both dead at this point) but did express his certainty that Pio prophesied the pontificate of Pope Paul VI.[22] Our own research has left us without the same degree of certainty, but it's entirely possible Pio could have predicted Paul VI's pontificate either because of supernatural prophecy if one is inclined to believe such things or through natural means through his knowledge of the workings of the Church. Prior to his election to the pontificate in 1963, Paul VI (then still known as Cardinal Giovanni Battista Enrico Antonio Maria Montini) was widely considered the most likely choice to be the new Pope due to his close working relationships with Popes Pius XII and John XXIII (whose work on the Second Vatican Council Paul VI would ultimately finish) and his administrative background.[23] That is to say, predicting who the next Pope will be is a similar exercise to predicting who will be the next politician to lead a country—there may not be certainty about it, but there are at least good reasons to suspect one person rather than another.

An interesting side note came to our attention when we were researching how hard it would be to predict who the next Pope might be: apparently bookmakers have gotten in the business of laying odds on the pontificate.[24] Apparently people will gamble on anything. Our guess is Padre Pio probably wouldn't have approved of that particular development, though Catholics are divided on the subject of whether gambling should be considered sinful or not. No way are we going to step into that debate.

20   Kwitny, J. (1997). *Man of the Century: The Life and Times of Pope John Paul II*. New York: Henry Holt & Co.

21   Dziwisz, S. (2007). *A Life with Karol: My Forty-Year Friendship with the Man Who Became Pope*. New York: Doubleday.

22   Amorth, G. (2016, 2021). [Trans: Sherry, M.]. *Padre Pio: Stories and Memories of My Mentor and Friend*. San Francisco, CA: Ignatius Press.

23   Duffy, E. (1997, 2014). *Saints and Sinners: A History of the Popes* (4th ed.). New Haven, CT: Yale University Press.

24   OLBG (2024). Next Pope Betting Odds. *OLBG*. <https://www.olbg.com/blogs/next-pope-betting-odds> (accessed July 1, 2024).

Returning to Padre Pio, one of the most often-repeated (yet poorly documented) cases involves the alleged miraculous healing of a young girl named Gemma di Giorgi. Born blind because her eyes lacked pupils, di Giorgi was said to have been given a miraculous gift of the ability to see despite the remaining deformity in her eyes after confessing to Padre Pio.[25] Though many sources have repeated the story, we have not found any primary documentation of the event nor results of any medical examinations that might shine light on the matter in either direction. All we have been able to determine is that Gemma is a real person, but we present the story of her alleged miracle without drawing conclusions.

The Vatican didn't always take a kindly view of the friar from Pietrelcina. Concerned that Pio was perhaps perpetrating a hoax to gain fame for himself (or, taking a more charitable view, to promote the faith), they ordered a number of investigations and imposed a number of restrictions on Pio. At one point, they attempted to relocate Padre Pio to a different parish where he was less well known, but violent protests from the local community forced them to allow him to stay in place. Likely the resulting controversy only made him even more popular among the faithful. Still, he was forbidden from hearing confessions or celebrating the Mass in public between 1931 and 1933, restrictions which were gradually lifted beginning July of 1933 and fully lifted by May of 1934.[26]

This rehabilitation was not permanent, but lasted long enough for the Padre to begin work to establish his Home for the Relief of Suffering, a research hospital founded by Pio and still administered by the Vatican to this day (the Home is considered by many the cornerstone of Pio's charitable works outside of his religious activities like hearing confessions or reading the Mass). But in 1952 the Vatican banned several books about Pio and by 1954 prevented him from tending to administrative duties at the Home. By 1958, he was under constant supervision, with secret microphones allegedly placed in his parlor.[27]

He was subjected to two Apostolic Visitations (essentially investigative visits by a superior in the Church hierarchy to ensure no abuses are taking place). The first of these was administered in 1921 by Bishop Raffaele Rossi whose report generally took a favorable view of Padre Pio, acknowledging the existence of the stigmata and proclaiming they were neither the work of evil nor of a deliberate hoax, though he also admitted he could not verify many of the miraculous events or healings attributed to Pio.[28] The second Apostolic Visitation was conducted in 1960 by Father Carlo Maccari which Father Amorth described as "disastrous."[29]

25   UCA News (2003). Visit of Woman Healed by Saint Padre Pio Inspires Many. *Union of Catholic Asian News.* <https://www.ucanews.com/story-archive/?post_name=/2003/03/28/visit-of-woman-healed-by-saint-padre-pio-inspires-many&post_id=22215> (accessed July 1, 2024).

26   Amorth, G. (2016, 2021). [Trans: Sherry, M.]. *Padre Pio: Stories and Memories of My Mentor and Friend.* San Francisco, CA: Ignatius Press.

27   *Ibid.*

28   Castelli, F. (2011) [trans: Messori, V., Bockhorn, L., & Bockhorn, G.]. *Padre Pio Under Investigation: The Secret Vatican Files.* San Francisco, CA: Ignatius Press.

29   Amorth, G. (2016, 2021). [Trans: Sherry, M.]. *Padre Pio: Stories and Memories of My Mentor and Friend.* San Francisco, CA: Ignatius Press.

Pio and Maccari did not get along and the latter's eventual report accused Pio of supporting "religious conceptions that oscillate between superstition and magic," said his supporters were "a vast and dangerous organization," and even mused as to how God could allow for such a level of deception.[30]

So who was right? Bishop Rossi or Father Maccari? Was Padre Pio a deeply religious man marked by a variety of supernatural occurrences or a dangerous charlatan? Because there is insufficient evidence to prove the case one way or the other, Rocky Mountain Paranormal maintains our strict neutrality. But the Church did eventually change its tune about Padre Pio. In the mid 1960s, Pio was fully rehabilitated by Pope Paul VI with whom he enjoyed a relationship of mutual respect.[31]

Even with the support of Pope Paul VI, how did Pio go from such suspicion to being not only made a saint but one of the most commonly venerated saints in recent memory? The process of attaining sainthood in the Catholic Church is not simple and has not remained the same throughout the entire Church history. For the early Church, sainthood was largely a matter of popularity. Individuals were venerated, some by the entire global Church and some in local communities, based largely on spontaneous veneration among the faithful public. Over the centuries, established formal procedures have been established by the Church to standardize the process. The current procedure has been in effect since 1983.[32] Though some cases take longer than others (and some progress further than others) for a variety of reasons, there are essentially four stages.

In the "Servant of God" stage, the local bishop initiates the proceedings with a preliminary investigation of the candidate's life. Assuming a favorable finding, the person is labeled a Servant of God and the case is forwarded to the Congregation for the Causes of Saints who then take over the process. Normally, this process must wait at least five years after the individual's death, but this requirement may be waived by the Pope.

If sufficient evidence to support the case is found by the Congregation for the Causes of Saints, the candidate may then be labeled "Venerable" in a Papal degree. These individuals are not given established feast days, but the faithful are encouraged to pray for their intercession as a sign of God's will that the candidate should move forward to be canonized as a saint.

The next step is called "Beatification," at which time the candidate is referred to as "Blessed." This again requires a Papal degree, and the process depends on whether or not the candidate was a martyr. For a martyr, the Pope simply declares their martyrdom. For a non-martyr, this stage requires evidence that a miracle has

30   Luzzatto, S. (2011). *Padre Pio: Miracles and Politics in a Secular Age*. New York: Picador.

31   Allen, J. L. (2001). For all who feel put upon by the Vatican: A new patron saint of Holy Rehabilitation. *National Catholic Reporter, 1*(18). <https://www.nationalcatholicreporter.org/word/word1228.htm> (accessed July 1, 2024).

32   John Paul II (1983). Apostolic Constitution: Divinus Perfectionis Magister. *The Holy See*. <https://www.vatican.va/content/john-paul-ii/en/apost_constitutions/documents/hf_jp-ii_apc_25011983_divinus-perfectionis-magister.html> (accessed July 1, 2024).

occurred through the candidate's intercession after their death. In either case, this is taken as evidence that it is "worthy of belief" that the individual is in Heaven. That is, the Church does not require that belief but does not prevent the faithful from holding it themselves.

The final step is "Canonization," at which time the individual becomes a "Saint." This is the Church's official declaration that the individual has attained the "Beatific Vision" in Heaven, and this recognition is definitively held by the Catholic faithful. Canonization as a Saint requires the evidence of two miracles attributed to the Blessed's intercession after their death (one in the case of martyrs; rarely, the requirement of a second miracle is waived by the uniform consensus of the Pope, the Congregation for the Causes of Saints, and the College of Cardinals).

Along the way, the cause for canonization is overseen by a "postulator" who is responsible for conducting the relevant investigations both into the individual's background and into allegations of miraculous occurrences. A "Promoter of the Faith," now more properly called the "Prelate Theologian"[33] fills the role colloquially known as the "devil's advocate" to oversee the proceedings and ensure no rash declarations are made.

Importantly, Catholic doctrine does not hold that the Saints themselves perform the miracles required for their canonization. Rather, they hold that God performs miracles through the intercession of the Saints. It's a subtle point of distinction and doesn't much matter when it comes to the scientific inquiry into the alleged miracles, but it is a key point of Catholic doctrine so we don't want to get it wrong. Catholics also hold that plenty of individuals are "saints" in the technical sense (meaning their souls are in Heaven) even if they've not been formally recognized as such by an official decree of the Church. Protestants (with the exception of just a few denominations), on the other hand, tend not to venerate individual saints, arguing instead that sainthood refers to all Christians (or at least all Christians who earned their eternal reward) and that veneration of individuals as saints is a form of idolatry. We're not here to settle theological debates, just to explain (as briefly as we can) what the different faiths believe.

Given this long and (especially to an outsider) seemingly convoluted process, how did Padre Pio end up becoming Saint Padre Pio? Popular calls for his sainthood began as soon as he died in 1968, but it took until 1982 for the Holy See to authorize the Archbishop of Manfredonia (in which diocese Pio lived and worked) to open an investigation and he was declared a Servant of God shortly thereafter and eventually declared Venerable by Pope John Paul II in 1997.[34]

The Congregation for the Causes of Saints recognized the miraculous healing of one Consiglia De Martino of Salerno in 1998, leading to his beatification by Pope John Paul II in 1999.[35] This miraculous healing involved a ruptured tho-

---

33   *Ibid.*

34   Amorth, G. (2016, 2021). [Trans: Sherry, M.]. *Padre Pio: Stories and Memories of My Mentor and Friend.* San Francisco, CA: Ignatius Press.

35   Vatican (n.d.). Padre Pio Da Petrelcina. *Vatican.* <https://www.vatican.va/news_services/liturgy/saints/ns_lit_doc_20020616_padre-pio_en.html> (accessed July 1, 2024).

racic duct which formed a lump containing some two liters of lymphatic fluid. It was supposed to require painful surgery to correct, but Consiglia says she prayed for Padre Pio's intercession after which the rupture seemed to correct itself. She attributed the cure to Padre Pio's intercession and the Vatican investigators determined that no scientific explanation could be found, allowing them to validate the miracle.[36]

Finally, Padre Pio was canonized by the same Pope John Paul II in 2002.[37] Some 300,000 people attended the ceremony at St. Peter's Square in Rome.[38] The miracle that supported this case involved a comatose young Italian boy named Matteo Pio Colella whose condition resulted from suspected meningitis. One day he simply woke up, entirely recovered, and said an old man with a white beard visited him in a dream and told him he'd recover. Of course, that dream-visitor is assumed to have been none other than Padre Pio.[39]

What are we to make of these kinds of miracles? Obviously we were not present to witness them, nor have we been given access to the medical files. What we can say is that there are different types and standards of evidence when it comes to science versus religion. What appears to a religious person as a miraculous cure might appear to a skeptical scientist simply as an unexplained medical phenomenon. Spontaneous recoveries or remissions of cancers do occur. Religiously predisposed individuals may attribute those to divine intervention. Others may attribute them to happenstance. This is the kind of thing that, in the absence of other evidence, we have to leave to the judgment of individual readers. These are the kinds of evidence that will simply be unconvincing to skeptics because the quality of evidence doesn't meet the burden of proof for scientific inquiry, but might meet the standard of evidence for belief as long as it's within the context of one's pre-existing religious philosophy. And since we didn't actively work on those cases, we don't want to step on anyone's toes.

A good analogy can be found within the doctrine we've already covered. In the "Blessed" stage of the canonization process, the Church considers the individual's presence to be "worthy of belief" rather than a proved or settled matter. Similarly, if a claim of a miracle—particularly something like a miraculous healing whose nature is by necessity somewhat "fuzzy"—can't be either debunked or proved, religious individuals may consider it "worthy of belief" based on their own faith, just as skeptics may consider it "worthy of disbelief"

36   Magis Center (2022). The Padre Pio Miracle That Led to His Beatification. *Magis Center.* <https://www.magiscenter.com/blog/the-padre-pio-miracle-that-led-to-his-beatification> (accessed July 1, 2024).

37   Vatican (n.d.). Padre Pio Da Petrelcina. *Vatican.* <https://www.vatican.va/news_services/liturgy/saints/ns_lit_doc_20020616_padre-pio_en.html> (accessed July 1, 2024).

38   Holley, D. (2002). Beloved Pio Becomes a Saint. *Los Angeles Times.* <https://www.latimes.com/archives/la-xpm-2002-jun-17-fg-padre17-story.html> (accessed July 1, 2024).

39   McCann, E. (2007). The other side of miraculous monk Padre Pio. *Belfast Telegraph.* <https://www.belfasttelegraph.co.uk/opinion/columnists/the-other-side-of-miraculous-monk-padre-pio/28478802.html> (accessed July 1, 2024).

based on their lack of faith. In these cases, we think the best thing to do is to just lay out as many facts as possible and then decline to take an official position on the resulting debate.

So that's the backstory. How exactly did Rocky Mountain Paranormal get involved? It all started with an online auction at a thrift store website which listed an old religious book without much detail as to its provenance. The listing identified it as a 1939 *Missale Romanum*[40] but gave no mention of Padre Pio himself. Because most of our members are devoted collectors of strange, bizarre, or historical artifacts, the member who first saw the listing took an interest in it. No one else seemed too keen on it and we didn't want to see this remarkable old book just get thrown away, so we bought it for somewhere in the neighborhood of $50.

When it arrived, we started to get a little more excited. Inside the covers of the book were inscriptions describing some of its history. Of particular note was an inscription reading "This is to testify that this book was used for the first time by Rev. Father Pio de Pietrelcina" and signed "On Corpus Christi of 1945, Father George Pogany." Now believing we might have accidentally just acquired a second class relic of Saint Padre Pio (he had already been canonized by the time we got the book), we started to dig into the history a little bit. Of course we always have to be careful lest we accidentally fall prey to some kind of hoax. However, since this wasn't advertised as having belonged to Pio, we thought it unlikely. Still, we wanted to establish its provenance and validate its history.

Our first step was to determine the identity of Father George Pogany. As it turns out, he was a priest who came from a Jewish family and worked as Padre Pio's correspondence secretary; even though the local community knew of his Jewish origins, they refused to betray him to Italian Fascists during the war.[41]

While that may not have been absolute confirmation of the book's authenticity, it at least elevated it to the level of "if not real, a very good hoax." We kept digging. Another inscription contained within the book said, at some greater length, "This missal was blessed and used by Padre Pio, Franciscan Monastery, San Giovanni Rotunda (Friggia), Italy, on or about the feast of the Sacred Heart, June 1945, and was presented to Saint Anthony Parish, San Bernadino, California by Mr. and Mrs. [redacted to protect the family's identity] and Family. It is the pious hope of the donors that Padre Pio and Father John St. John, S. J.[42], as well as the donors be remembered whenever this missal is used at the altar. Given to the first pastor of Saint Anthony's Church, Michael J. Brown on the ground breaking day and the First Holy Communion of the children, the Feast of Saint Anthony, June 13, 1948."

At this point, we had some further information and some further questions.

---

40   The Roman Missal is the book Catholics use to read the mass. Lay faithful may purchase individual copies of the Roman Missal with which to follow along with the readings at the Mass. Clergy use much larger and more specialized versions from which they read when they celebrate the Mass.

41   Pogany, E. L. (2001). *In My Brother's Image: Twin Brothers Separated by Faith After the Holocaust.* New York: Penguin.

42   The "S. J." postnominal indicates membership in the Society of Jesus, better known as the Jesuits.

We were first able to determine that indeed there is a Saint Anthony parish in San Bernadino, and indeed it was founded by Father Michael Brown in 1948.[43]

We now believed we were in possession of a genuine second class relic of Padre Pio. We reached out to several Padre Pio societies for assistance in validating the item. They were able to confirm that both Father Pogany and Father St. John often had contact with Padre Pio in Italy, Pogany as his secretary and St. John as a US Army Chaplain who served part of his time in Italy.

Though the Padre Pio societies were able to confirm the authenticity of the book (even calling it a "treasure" in their correspondence with us), there are a couple gaps in the book's provenance we have to fill in with speculation. It's not entirely clear how the book passed from Padre Pio to the family who eventually donated it to the Saint Anthony Parish. Given their intention that both Padre Pio and Father St. John should be remembered during its use, we believe they were acquaintances of Father St. John. Either Padre Pio gave the book to Father St. John who subsequently gave it to this family or, perhaps a bit less likely, this family made the acquaintance of Padre Pio through Father St. John and received the book in recognition of some gift or service they performed for the Friary. Exchanges of such gifts are not uncommon in the Catholic Church. Indeed, this practice is even evidenced by the family's donation of the Missal to Father Brown and the Saint Anthony Parish.[44]

The other gap in the book's provenance is how it went from Saint Anthony's to the thrift store auction. We were able to determine that it had not been donated by the parish itself, so we believe it was given by Saint Anthony's to a member of the parish, probably also in thanks for some service offered. Churches tend not to pay people for their service but when someone donates something substantial or performs service above and beyond what's expected, gifts of some sentimental value are commonplace. Most likely, the individual who received the book kept it tucked away as a treasure until that individual's estate, not understanding what it was, eventually donated it to the thrift store.

Despite these couple of gaps in the provenance, we're confident in the book's authenticity. Its inscription includes details that are not widely known. For instance, until we contacted the Padre Pio society, we were unaware of such a character as Father John St. John (and even upon learning of his acquaintance with Padre Pio, have not found any readily available documents mentioning both of these two men). Between our own research and the authentication from the Padre Pio experts, we were prepared to call it an authentic second class relic of Saint Pio.

And then we noticed something even more remarkable. Some of the pages had unusual small brown stains. Closer inspection confirmed these to be blood stains. Recall that Padre Pio, for most of his life, had the stigmata on his hands.

---

43   Diocese of San Bernadino (n.d.). Parish History: History of St. Anthony. *Diocese of San Bernadino.* <https://www.sbdiocese.org/parishes/history.cfm?id=17285> (accessed July 1, 2024).

44   Many priests, sometimes even entire parishes, don't have a lot of personal resources and rely on gifts from parishioners to supply the materials they need to carry out their clerical duties.

Though he often kept his hands bandaged or covered with gloves, he had to remove the coverings to celebrate the Mass. What we initially thought was a second class relic of Saint Pio turned out to contain blood stains from the Saint, making it a first class relic.[45]

Once we made our discovery, the Church asked us to donate the book back to them. We certainly understand why, but we assured them that, while we had (and have) every intention of keeping it in our own archives, it is very well cared for. On very rare occasions, we even bring it from its permanent home containing our unusual archives to exhibit it for the public. And of course because touching an item to a first class relic makes that item into a third class relic, we obtained a bunch of Saint Padre Pio medals and created some new relics of our own.

So that's how an odd little paranormal investigation group from Colorado ended up in the possession of, and playing a role in the authentication of, a first class Catholic relic of one of the most highly venerated Saints in recent memory.

An interesting addendum has come up a few times over the years. Clearly we collect a lot of weird things. Our archives include, in addition to a wide variety of historic artifacts, horror movie props, and other oddities, plenty of items recovered (with permission, of course) from allegedly haunted places. We have items said to house dangerous curses. Sometimes when people visit our collections, they remark how frightening it must be to spend so much time around these objects and they often ask whether we've experienced any hauntings or negative occurrences because of these things. Honestly, we have to say no. We'd even welcome someone or some thing to haunt our houses because that would give us the kind of evidence and experience we've been looking for, but it's never happened for us. The skeptical reason, and the one we've always found most likely, is that, quite simply, nothing has ever attached itself to any of these objects and followed us home. But on a couple of occasions some people have suggested that maybe we've been protected by divine providence because our archives are also home to Padre Pio's book.

It's not the most parsimonious explanation, but it certainly is a fun one. And as much as we always insist on good scientific answers, sometimes it's important to think of the fun answers at the same time.

---

45   At least in the old system of classifying relics still most commonly observed by the Catholic faithful.

Afterword

Well, here we are at the end of yet another book. What an adventure this has been! And we mean that in every context possible. On one level, the adventures were the cases themselves. We've been investigating strange, bizarre, and paranormal claims of all varieties for over twenty-five years now. Each of those investigations represents a unique and wonderful adventure. Part of the pleasure in this work is that it affords us the opportunity to see places other people don't get to see, or at least to experience them in a context other people don't get to experience them. Who else gets to experience the (not open to the public) archives of museums, recreate ghost and alien videos, poke around in magnificent historic mansions, and do it all in the name of science?

But writing these books is an entirely different kind of adventure. Any literary endeavor is a monumental undertaking. To write a book requires dedicated attention over an extended period of time. That's true of any book, but in the case of these books, we get to add the additional pleasure and challenge of writing about history. It's a challenge because it often involves going down dead-end rabbit holes trying to suss out the origin of some unsubstantiated rumor or another. But it's also a genuine pleasure because it gives us the chance to share stories that might otherwise be forgotten. We've long mourned the decline of historical literacy in our culture, and if some good old fashioned ghost stories are what it takes to get people to read history again, so be it. And so much the better because it means we get to play in a field (historic research) near and dear to us.

A third kind of adventure has emerged since the publication of our previous volume. That was the first book Rocky Mountain Paranormal ever wrote and it launched us into the exciting world of publishing, book signings, and literary events. Fortunately, it's been well-received. It seems like there actually is an audience out there for our unique brand of paranormal investigation that balances the haunting with the historic, the spooky with the scientific, and the frightening with the forensic. We genuinely hope readers enjoyed this second volume as much or more than (they told us) they enjoyed the first one. If that's the case, never to fear: by the time anyone reads these words, we'll already be hard at work on Volume 3, and we don't show any signs of stopping.

Even as we write these words, brand new cases are consistently flowing into our files. Some of them will never come to anything, while others may be destined for publication in Volumes 4 and beyond. Whatever your opinion of the paranormal may be—and we think we're almost unique in that our work seems to appeal equally to believers and skeptics, so we assume there's a good mix of such philosophies in our readership—one thing is absolutely clear: belief and interest in these topics is still going strong. Sure, the details may shift from time to time. Sometimes people are more interested in ghosts; other times, they want to read about aliens. But the foundational interest in paranormal subjects is always there. And wherever there's some high-quality weird, we'll also be there, trying to figure it all out.

An old adage reminds us that one should never discuss politics or religion in mixed company. So of course we ended up closing this volume with chapters that dealt explicitly with a political campaign and a religious figure. It seemed like a good idea at the time, and we hope readers take us at our word that we're not arguing for or against any political or religious ideas except to the limited extent directly related to our work to document and investigate paranormal claims and then to explain and educate the public about our findings.

But if you thought those activities represent the weirdest of our case files, you haven't seen the half of it yet. We haven't even gotten to our most terrifying encounter (at a location that was the site of tens of thousands of deaths), our encounter with a cricket demon, the time one of our members almost, quite by accident, developed a religious following by staring at people, the "exorcism by breakfast cereal" affair, and dozens of other fascinating cases we have slated for publication in future volumes. Be sure to stay tuned for those, and if you haven't read our first volume yet, be sure to check that out in the meantime.

Until next time, then, thanks for reading and stay spooky!

If you've enjoyed this book, don't forget to leave us a review online. Reviews of even just a few words are immeasurably helpful to independent authors and publishers. And the ego boost will help get us through the process of writing Volume 3.

Acknowledgements

Though writing a book is often seen as a solitary endeavor, it necessarily involves the help and support of numerous people. That's doubly the case when the book describes real-world case files spanning a quarter of a century. As such, the number of people to whom we owe our sincere thanks is beyond our ability to count, and any attempt at a complete listing would be doomed to failure. Nevertheless, we'd like to begin by offering our sincere thanks to all of those who've allowed us to conduct our investigations in their homes or businesses over the years. Without them, this book clearly would never have happened. Similarly, we'd like to thank everyone who has joined us on investigations over the years. Both groups of people would require another full-length book to list, so we thank them all collectively. We would like to specifically highlight the ongoing support and hard work of Carol Olivacz, Jack Hanley, and Kathy Josey, who've been tirelessly working with us for years, as well as all the former members of Rocky Mountain Paranormal.

Special thanks are due to the entire crew at the Colorado Festival of Horror. Their support has been immeasurable, not only as we prepared this book but as we've delivered our lectures and more over the years. And readers are encouraged to attend their annual festival and show them some love.

Thanks also to Arturo Spraycasso for the truly excellent interior artwork that helps bring some of these stories to life.

Finally, thanks to everyone who bought and read a copy of this book and/or its predecessor. Without you, none of this would mean anything.

About the Authors

Robert (Bob) Lewis is a Colorado-based author, editor, paranormal investigator, scholar, magician, and more. He holds degrees from the University of Colorado Denver in Biology, English, Mathematics, and Psychology, and a Master of Education from the University at Buffalo in Science and the Public. A dedicated polymath, he likes to tell people that his hobby is to collect new hobbies. He's (obviously) a member of the Rocky Mountain Paranormal Research Society. Additionally, he's a co-host of the *Do You Like Scary Movies* horror podcast, host of the *Phobophile* YouTube channel, and is always looking for more projects to whittle away at what little time he has left for sleep.

In addition to the first volume of this series, his publications include *In the Woods: A Fiction Foundry Anthology* and *Arithmophobia: An Anthology of Mathematical Horror,* both of which are available from Polymath Press. He can be found online at www.robertlewisauthor.com.

Bryan Bonner is a founding member of the Rocky Mountain Paranormal Research Society. For over two decades, Bryan has dedicated himself to the examination of a wide range of paranormal phenomena, including ghosts, poltergeists, psychics, UFOs, conspiracy theories, urban legends, and much more. Setting himself apart from others in the field, Bryan has always maintained a grounded approach, refraining from running around cemeteries at night and needlessly scaring others with imaginative tales. Instead, he meticulously tests bizarre beliefs and practices, conducts experiments and on-site investigations, and even recreates unusual events, all with the aim of uncovering the truth. Bryan's work has garnered the respect of believers and skeptics alike, while simultaneously instilling fear in fraudsters and charlatans. You can read more about Bryan and his work at www.rockymountainparanormal.com.

About the Artist

Arturo Spraycasso, a celebrated artist known for his mastery of spray paint, has made significant strides in the art world since bursting onto the scene in 1998. Representing New Mexico at expositions across the United States and excelling in various media, Spraycasso found his true calling in aerosol art. His innovative techniques and tools have revolutionized spray painting, earning him widespread acclaim and a successful TV series, *Spraycasso: The Art of Spraypaint*. Recently, he has expanded his artistic expression into digital media. This transition has not only diversified his work but also enhanced its accessibility and versatility. Although he typically focuses on landscape and mystical themes, his new digital creations for a supernatural book project showcase his adaptability and boundless creativity. His work, both traditional and digital, continues to captivate audiences in galleries across Colorado and New Mexico. Learn more about Spraycasso and his work at www.spraycasso.com.

Also available from Polymath Press

Case Files of the Rocky Mountain Paranormal Research Society Volume 1 by Robert Lewis & Bryan Bonner

Don't miss the first volume of the *Case Files of the Rocky Mountain Paranormal Research Society* series!

Included in this volume:

*How Rocky Mountain Paranormal involved multiple departments of the United States Federal government in a paranormal investigation
*The haunting of an active nuclear military base
*A paranormal investigation inside a jail
*A paranormal investigation deep inside a cave
*A location that is still home to thousands of unidentified human bodies—and how Rocky Mountain Paranormal worked to tell their stories
*Colorado's own vampire legend
*The time one of our investigators got slapped by an unseen entity
*The inspiration behind a classic horror movie
*And many more!

In the Woods: A Fiction Foundry Anthology
Edited by Robert Lewis

Strange things can happen in the woods.
Sometimes they're frightening.
Sometimes they're funny.
Sometimes they're just plain weird.

The authors of the Fiction Foundry writers' critique group have taken it upon themselves to explore all the strange things that happen in the often majestic and yet often harsh woodlands.

Fiction Foundry, established 2012, is a group of writers dedicated to helping prepare one another's work for professional publication. In this anthology, the group's authors show off their eclectic visions of life among the trees.

Featuring contributions by John H. Howard, Sangita Kalarickal, Josh Snider, Carolyn Kay, Robert Lewis, Charli Cowan, Henry Snider, Shiloh Silveira, Kari J. Wolfe, Christophe Maso, and Hollie Snider, this anthology brings us out of urban life and shows a world of forest spirits, haints, mental illness, parasitic spiders, werewolves, out of control plants, evil forces, reincarnation, humans with animal ears, witches, and Lovecraftian horrors.

And all of them can be found...*In the Woods*.

Arithmophobia: An Anthology of Mathematical Horror
Edited by Robert Lewis

"Arithmophobia," *n*.: The fear of numbers or mathematics.

Whether you love mathematics or find it terrifying, this anthology of original tales of terror is sure to send a chill down your spine. With an unlucky thirteen brand new horror stories and a bonus poem in case any readers suffer from triskaidekaphobia, these pages combine the talents of some of the genre's most experienced award-winning practitioners of terror and some of the literary world's most promising new voices.

Featuring contributions by Elizabeth Massie, Miguel Fliguer, Mike Slater, Patrick Freivald, Liz Kaufman, Damon Nomad, Sarah Lazarz, Martin Zeigler, Josh Snider, Rivka Crowbourne, Joe Stout, Brian Knight, Wil Forbis, David Lee Summers, and Maxwell I. Gold.

These stories tell us of strange and horrifying new geometries, crazed and violent mathematicians, sentient and malevolent numbers, and even some new mathematical twists on some classic monsters. You needn't be a mathematician to experience these new forms of mathematical terror, though students of the discipline might recognize some familiar names and ideas lurking in the shadows.

So pull up a chair, dust off your abacus and slide rule, and prepare to experience…

Arithmophobia.

Get your copies of these and other Polymath Press
titles online at
www.polymathpress.com
or wherever fine books are sold!